T0346812

# PRAISE FOR *KOJI ALCHEMY*

"They say a gram of koji contains millions of spores. Shih and Umansky equal that with insights, bringing scientific understanding to koji's magical powers of transformation (without spoiling any of the magic). *Koji Alchemy* is an immensely informative read."

—**DAN BARBER**, chef/co-owner, Blue Hill and Blue Hill at Stone Barns; author of *The Third Plate*

"I have always been intrigued by the technical and thoughtful fermentation practices of many cultures. Reading *Koji Alchemy* has opened my eyes to how simple working with koji can actually be, while remaining incredibly innovative and exciting. This book is not only full of information, it's scientific, and most of all thrilling. It's a great reminder that we all have a lot to learn about the art of fermentation."

—**DANIEL BOULUD**, chef and restaurateur, Daniel

"Before this book, what Jeremy and Rich have done with food I would've thought impossible. As someone who has always had a deep love for the curing, brining, aging, and smoking of meat, I would have never thought these time-honored processes could be sped up without producing lesser results. Well, was I ever proven wrong. To take an ingredient like koji, which has been available and used for hundreds of years, and use it to speed up processes like pastrami and still maintain its integrity is truly astonishing. In *Koji Alchemy*, they show you what to me feels like magic, and guide you through every step so you can now make dishes at home or at your restaurants that would have previously taken not only lots of time but also space. They have opened my mind to koji's limitless uses and will do the same for you. Jeremy and Rich, thank you for always pushing culinary boundaries and reminding us all that there is something new to learn and be inspired by every day."

—**CHEF MICHAEL SYMON**

"Science is a very useful resource for understanding food, and, very importantly, it shows us how to make better food. This book succeeds in 'translating' the language of science and its practical applications to everyday users. Exploiting and harnessing the effects that microorganisms have on food is a great tool that any cook can add to their repertoire, and *Koji Alchemy* will show you how. I hope Merriam-Webster adds the word 'kojify' to the dictionary soon."

—**FRANCISCO MIGOYA**, head chef, Modernist Cuisine

"*Koji Alchemy* is the most in-depth study on the subject that I've ever seen. As someone basically starting at zero, I'm inspired to dig in and learn from the masters."

—**JEREMY FOX**, author of *On Vegetables*

"Rich and Jeremy have dedicated years to researching and experimenting with both traditional and utterly novel uses for koji. With *Koji Alchemy*, they open their notebooks to the rest of us, sharing their deep knowledge and infectious enthusiasm for this remarkable mold. This book, at once comprehensive and approachable, will prove invaluable to any curious cook looking to make more flavorful food, reduce kitchen waste, or experiment at the frontier of an exciting world of fermentation."

—**DAN SOUZA**, editor-in-chief, *Cook's Illustrated*

"Alchemist, chef, and master charcutier, Jeremy Umansky is the Albert Einstein of koji. *Koji Alchemy*—his book with Rich Shih, a first-rate koji explorer—is fascinating, hunger-inducing, and the final word on the spore responsible for miso, soy sauce, sake, and so much more."

—**STEVEN RAICHLEN**, host, PBS's *Project Fire* and *Project Smoke*

"Like a magician revealing their tricks, *Koji Alchemy* has lifted the veil on the complex nature of some of the world's most beloved foods. The authors brilliantly demystify the 'alchemy,' giving the reader a straightforward view into the world of koji's secrets. Never before has a book broken down the building blocks of koji and its products so completely and in such a useful way in the kitchen."

—**KYLE CONNAUGHTON**, chef and owner, SingleThread Farms

"Rich and Jeremy have done an amazing job in creating an indispensable manual for both home cooks and pro chefs. *Koji Alchemy* offers a fresh set of transformative cooking fundamentals and a master class on fermentation."

—**RICK TRAMONTO**, executive chef and vice president of culinary operations, Tramonto Cuisine Group

"It's not every day that deeply passionate, creative, and collaborative people take on the demystification of a revered ingredient. But herein lies just that: a truly educational, comprehensive, and delicious exploration of one of the most versatile molds out there. So grateful for this book."

—**CORTNEY BURNS**, chef; author of *Nourish Me Home* and coauthor of *Bar Tartine*

"*Koji Alchemy* will make you want to immediately start experimenting. The fungus behind koji—*Aspergillus oryzae*—is one of the top three most important transformative organisms in food, and this is the only book devoted to it. Shih and Umansky have put in years of intensive learning and testing of all aspects, both traditional and new, of this magical mold and have generously shared their insights and given us a roadmap to follow."

—**DAVE ARNOLD**, author of *Liquid Intelligence*; Museum of Food and Drink

"*Koji Alchemy* is ideally suited to this moment in food history when the nutritive and ecological benefits of fermentation as a means of processing food to greater edibility have become baldly apparent. This is a handbook explaining how the distinctive enzymatic effects of koji, a mold domesticated in Asia in deep antiquity, produce richer flavor (more than just umami), easier digestibility, and more beneficial interactions with gut microbes than other cooking agents. You learn how to grow it, apply it, and think about the possible interactions with different grains and starches. The authors explain how koji performs its transformations in all of the traditional Asian categories of fermented food—the amino pastes, alcohols, meat cures, and pickles—then venture into transformations of eggs and dairy and western foodstuffs. More an invitation to work with koji to blaze creative paths in cooking than orders about how to ferment using *Aspergillus oryzae*, this book stresses the wonder of this fungus's ability to give rise to subtle fragrances, comforting tastes, and a feeling of wholesome repleteness in food."

—**DAVID S. SHIELDS**, Carolina Distinguished Professor, University of South Carolina; chair, Carolina Gold Rice Foundation

"*Koji Alchemy* opens up a whole new world of endless gastronomic possibilities. Koji will be the next great wave for fermentation enthusiasts, and these authors are its heralds. I have watched from the sidelines with great anticipation and am now ready to dive in myself with this book as my guide."

—**KEN ALBALA**, professor of history, University of the Pacific

"In *Koji Alchemy*, Umansky and Shih have amassed and arranged koji's mystifying powers into an easily digestible reference for the whole world to use! Both seasoned veterans in the field of fermentation, they speak to enchanted newcomers and inquisitive professionals with the certain euphoria that *Aspergillus oryzae* has long-deserved. *Koji Alchemy* gives praise to past and present traditions, while bringing modern concepts forward. This book, like koji itself, will turn the confounding into commonplace, blandness into flavorsome, emboldening your appetite for more umami."

—**MICHAEL HARLAN TURKELL**, author of *Acid Trip*

"*Koji Alchemy* is about discovery. It is about collaboration with microbes and humans. It is about cultural exchange, and most importantly, it is about making flavor. Jeremy and Rich gracefully navigate this ancient art and its modern renaissance that breaks from tradition, bringing us a complete look at this versatile domesticated fungus."

—**KIRSTEN K. SHOCKEY**, coauthor of *Miso, Tempeh, Natto and Other Tasty Ferments* and *Fermented Vegetables*

"*Koji Alchemy* is part science experiment, part history book, part love letter to the art and skill of using koji to elevate food both humble and refined. Shih and Umansky make the world of koji accessible to cooks at every level with this in-depth manual that shares their immense knowledge and passion with readers."

—**LEE WOLEN**, chef and partner, Boka and Somerset restaurants

"Rich and Jeremy have dedicated their lives to the art of making koji, and it shows very evidently in their work. *Koji Alchemy* is one of the most comprehensive pieces of literature regarding this amazing mold in the industry today. This book is a tool every kitchen should be using."

—**CHEF RYAN POLI**

"In an ever-evolving culinary world, *Koji Alchemy* will help define the next generation of cooks and eventually become a timeless resource for both those who are beginning their journey and those looking to add and improve on an existing skill set. An ancient technique broken down for the modern age. Old school flavor, new school style."

—**CARLO G. LAMAGNA**, chef and owner, Magna Kusina

"When I first met and became friends with Jeremy Umansky in 2014, he exposed me to what he had been working on with regard to koji. It was the kind of culinary awakening that opened my eyes to what is possible with food in order to create deliciousness, and this has carried over into everything that I do in the kitchen to this day. I couldn't be more excited for the knowledge in his and Rich Shih's beautiful book to be widely available. *Koji Alchemy* will change you."

—**KEVIN SOUSA**, chef and owner, Superior Motors

"What an extraordinary book *Koji Alchemy* is. I leapt headfirst into the section on vinegar, but from there couldn't resist discovering all the other ways that Shih and Umansky expound on the flavour impact of koji. It's a book for ingredient geeks, for sure (which I mean as a compliment), but written engagingly enough to appeal to anyone with a serious interest in food. My culinary mind is blown."

—**ANGELA CLUTTON**, author of *The Vinegar Cupboard*

"*Koji Alchemy* is a great, well-written book that draws you in from the start, equipping readers with the knowledge needed to discover more about this mysterious ingredient. Koji is well-documented as a core ingredient in Japanese cooking, a starting point with plenty more to be discovered. I love that this book demonstrates how its use for imparting and discovering flavour can be continuously explored in all cuisines. *Koji Alchemy* is the road map that sets you on that journey."

—**YUKI GOMI**, Yuki's Kitchen; author of *Sushi at Home*

"*Koji Alchemy* dissolves the boundaries of traditional koji methods, while still heeding the wisdom and power of cultures who have collaborated with these microbes to develop unparalleled flavors. Put koji on it."

—**LINDSAY WHITEAKER**, Harvest Roots Ferments

# KOJI Alchemy

## Rediscovering the Magic of Mold-Based Fermentation

**RICH SHIH** *and*
**JEREMY UMANSKY**

*Foreword by* **SANDOR ELLIX KATZ**

CHELSEA GREEN PUBLISHING
WHITE RIVER JUNCTION, VERMONT
LONDON, UK

Project Manager: Patricia Stone
Developmental Editor: Makenna Goodman
Project Editor: Michael Metivier
Copy Editor: Laura Jorstad
Proofreader: Angela Boyle
Indexer: Shana Milkie
Designer: Melissa Jacobson

Printed in Canada.
First printing April 2020.
10 9 8 7 6 5 4    24 25 26 27 28

DISCLAIMER: Information offered in this book is based on years of experimentation, experience, and research. Parameters for safety and warnings of dangers are presented throughout the book and ought to be heeded. However, the authors are not trained professionals in food science, food safety, health care, or microbiology; neither they nor the publisher are responsible for the consequences of the application or misapplication of any information or ideas presented herein.

**Our Commitment to Green Publishing**
Chelsea Green sees publishing as a tool for cultural change and ecological stewardship. We strive to align our book manufacturing practices with our editorial mission and to reduce the impact of our business enterprise in the environment. We print our books using vegetable-based inks whenever possible. This book may cost slightly more because it was printed on paper that contains recycled fiber, and we hope you'll agree that it's worth it. *Koji Alchemy* was printed on paper supplied by Marquis that is made of recycled materials and other controlled sources.

**Library of Congress Cataloging-in-Publication Data**
Names: Shih, Rich, author. | Umansky, Jeremy, author. | Katz, Sandor Ellix, 1962– writer of foreword.
Title: Koji alchemy : rediscovering the magic of mold-based fermentation / Rich Shih and Jeremy Umansky ; foreword by Sandor Ellix Katz.
Description: White River Junction, Vermont : Chelsea Green Publishing, 2020. | Includes bibliographical references and index.
Identifiers: LCCN 2019057468 (print) | LCCN 2019057469 (ebook) | ISBN 9781603588683 (hardcover) | ISBN 9781603588690 (ebook)
Subjects: LCSH: Fermented foods. | Cooking (Fermented foods) | LCGFT: Cookbooks.
Classification: LCC TX827.5 .S55 2020 (print) | LCC TX827.5 (ebook) | DDC 664/.024—dc23
LC record available at https://lccn.loc.gov/2019057468
LC ebook record available at https://lccn.loc.gov/2019057469

Chelsea Green Publishing
White River Junction, Vermont, USA
London, UK
www.chelseagreen.com

To our wives who understand that
we can't help but pursue our passions, and
to our daughters whom we hope will
understand them when they're old enough.

# CONTENTS

# FOREWORD

Koji is a mold with great transformative powers. It is *Aspergillus oryzae* along with a few other related fungi, which, when grown on grains or other nutritive substrates, produce a treasure trove of digestive enzymes. Koji, in many varied forms, has been used by humans for thousands of years. In sake and other alcoholic beverages produced across Asia from rice, other grains, and starchy tubers, koji's amylase enzymes break down complex carbohydrates into simple sugars that yeast can ferment into alcohol. In soy sauce, miso, and other fermented pastes and sauces, koji's protease enzymes break down proteins into amino acids, including glutamates with powerful umami flavor. Koji also has traditional applications in vegetable pickling, sweets, fish sauce, and undoubtedly other foods and drinks that I have not seen or heard about. But as koji and its equivalents in other cultures have become more known beyond the lands where they are traditional, and as interest in fermentation has been experiencing a broad revival, experimentation has exploded, and exciting new applications of koji abound.

For the first decade of my fermentation journey, even after the publication of my first book, *Wild Fermentation*, in 2003, I was intimidated by the prospect of making koji. From *The Book of Miso* by William Shurtleff and Akiko Aoyagi, I know that growing koji from spores requires up to forty-eight hours of warm, humid conditions. I had made lots of tempeh using an oven with a pilot light to maintain similar conditions, but only for twenty-four hours, which was plenty challenging. Because the oven pilot slightly overheated the oven if the door was tightly shut, I generally left it ajar and needed to monitor the temperature and adjust the opening every few hours. Also, I lived in a community where I shared a kitchen with a dozen or more people, and commandeering the oven for two days seemed likely to inconvenience people. And because we were living off the electrical grid with a somewhat unreliable solar system, I was reluctant to use electrical means of regulating temperature.

I've been working with purchased koji since 1993, when my burgeoning interest in fermentation led me to my first experiments making miso and amazake. When I called the American Miso Company in North Carolina (which has since become Miso Master) to ask if they would sell me some koji, they were so amused I was making miso that they sent me that first koji free of charge, as a gift for the novelty of making my own miso. As miso making became an annual ritual in my life, I continued to purchase koji from American Miso Company and later from South River Miso Company in Massachusetts.

Finally, around 2005, I decided I was ready to try making my own koji. I ordered spores from GEM Cultures. I soaked barley and steamed it. I cooled the steamed barley to body temperature, introduced the spores, distributed them, transferred the inoculated barley into a cloth-lined hotel pan, and incubated it in the oven with the pilot light on. By the next morning the whole kitchen was engulfed in the sweet aroma of the koji, the mold seducing me with its intoxicating smell and the promise of incredible flavors to come.

Over the years I figured out other incubation systems that were more self-regulating and less inconveniencing. I wrote about making koji in *The Art of Fermentation* and have made koji with hundreds of people in my workshops. Ever since koji making came into my life, it has been a special event I always look forward to, a ritual marked by the sweet aroma that overtakes the whole house. Koji making has also become something I love to share, to demystify the process while enabling people to experience its magic.

I have grown koji on pearled barley more than on any other substrate. It's so delicious and so consistently easy, meaning that steamed pearled barley has enough moisture for the mold to develop quickly and luxuriantly every time without fuss. I have also made koji on different types of rice, millet, soybeans (with and without wheat), and fava beans (with and without wheat), and have even experimented with growing koji on sweet potatoes and chestnuts, both of which I grow. The major way I've mixed it up in miso making has been to replace soybeans with different beans such as chickpeas, limas, pintos, cowpeas, lentils, and mayocoba beans. But although I like to experiment, in the end I possess a pretty limited culinary imagination.

In my travels I've encountered people with extraordinary culinary imaginations who experiment with very exciting, visionary uses of koji. I loved Momofuku's incredible pistachio and pine nut misos. At the Noma test kitchen in Copenhagen, I tasted rich, delicious koji-digested grasshopper "garum." In Boston, Rich Shih collaborated on a fermentation-themed meal that I participated in, and blew my mind with popcorn koji. And via social media I watched as chef Jeremy Umansky—whom I had met and taught when he was a culinary student—documented his wild experiments applying koji to meat curing, vegetable curing, and other unprecedented applications.

Shih and Umansky's *Koji Alchemy* is a book that lives up to its title. Koji is extremely versatile and has varied applications, many as yet unimagined. The authors are both mad experimentalists, and *Koji Alchemy* features more ways of using koji than I have ever seen or heard of before, including groundbreaking ideas for growing koji on cured vegetables in order to create vegetable charcuterie with incredible flavor and texture; koji cultured butter and cheeses, as well as cheese miso; and much, much more—even puffed koji for crispy rice treats. But at every juncture in the book, after explaining their methods, and letting some of their featured collaborators explain theirs, Jeremy and Rich offer ideas for how you might vary the process. They encourage further exploration and personal alchemy, working with the fungus and developing a relationship with it, as you, and we all, learn how very versatile it is.

*Koji Alchemy* already has me experimenting with things I wouldn't have thought I could do. It is empowering and inspiring, and does much to open the door to further creativity and innovation. Throughout the book Jeremy and Rich repeat the mantra and hashtag #kojibuildscommunity. It does, and this book will surely expand the koji making community. I can't wait to see and taste some of the next wave of koji experimentation that it is likely to inspire.

—Sandor Ellix Katz

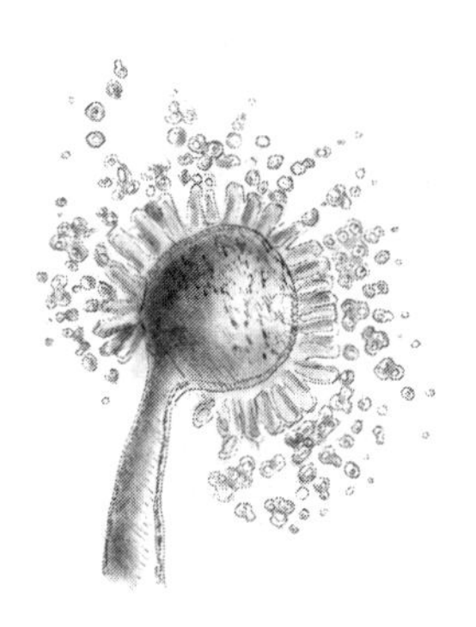

INTRODUCTION

# Why Koji?

There was a time when sitting down with family and friends at a table to share a meal was commonplace. It was a communal table where you could enjoy others' company without feeling like you had to do anything else; a time you looked forward to because you were welcomed by nourishment and people who cared; a daily forum to talk about anything you liked—from who you met that day to the deepest personal revelation; a place where you were comfortable sharing your emotions knowing that even if things got heated with disagreement, at the core everyone sitting at the table had your back. This may seem a romantic notion for most of us, but it's not. This is where we're headed with this book, so pull up a chair and join us.

Yes, it's become rare to sit at a table and enjoy a meal together. Most times folks are engaged in looking at a screen while wolfing down something out of a take-out container or plastic wrapper more for fuel than for joy. Somehow the industrial and green revolutions convinced many of us that cooking a meal is a burden. But there's a community starving for something more: the amazing food revolution—still in its infancy—of people who are fermenting vegetables (sauerkraut, kimchi, hot sauce, and the like), making their own charcuterie, raising chickens, learning whole-animal butchery, and foraging. An interest in koji has been running alongside this fermentation fever and is becoming more and more mainstream. We're here to show you the significance of this food revolution.

*Koji* is a specific mold grown on a starch medium that can make food extraordinarily tasty if you simply mix a few ingredients together and wait. *Alchemy* is a power or process that changes or transforms something in a mysterious or impressive way. Given koji's almost mystical abilities to create what seems like something out of nothing, we decided to name this book *Koji Alchemy*.

As you begin your journey with us, you may not know what koji is, but you're intrigued by what you've heard it can do. Maybe you've been told that it smells like the most intoxicating combination of grapefruit, chestnuts, honeysuckle, and mushrooms. It may sound hard to believe, and you want to taste what's actually possible. Well, it's likely you already have. You probably even have a bottle of it in your pantry right now. We're talking about soy sauce, a widely enjoyed condiment that is part of the common culinary language practically everywhere in the world, a familiar friend and a known tasty quantity that you've been dipping dumplings and sushi in longer than you can remember. Let's not forget the countless splashes you've added to marinades as a tried-and-true accent to make whatever you grill, sauté, or roast shine. As delicious as soy sauce is, we'll show you that it only scratches the surface of what can be done with koji.

To help make it easier to understand in application, we like to describe koji as a seasoning. Consider salt—the most basic seasoning that makes everything taste better. We've all experienced food that lacks the right amount—and we also quickly understand when there's too much. Think about a tomato at peak ripeness to be sliced and eaten. Now remember when you tasted the mind-blowing flavor of that tomato when you sprinkled a touch of salt over it; it's possible you can't imagine eating it any other way. The same goes for the primary driver of koji, *umami* (flavor body). Have you ever tried making a sauce *without* the bits of meat stuck to the pan after searing, aged cheese, mushrooms, tomato, seaweed, miso, or the like? The results will be missing depth of flavor, one that cannot be realized without amino acids, the fundamental components of proteins.

We are wired to taste umami and feel satisfaction, because it's an indicator of nutrition. When food lacks this savory taste, our bodies tell us it's not fulfilling. That's why we simmer water with bones to create a stock that ultimately makes a soup more satisfying, which doesn't take much protein to achieve. However, the ability to coax the flavor and nutrition out of protein requires cooking skills and/or preservation methods driven by heat, microbes, and time. The magic of koji is its ability to give us delicious foods with less effort and time. In some cases it works overnight. Simply applying koji to foods ahead of cooking leverages enzymes (which will be described in detail later) to accelerate the flavor generation process; the food is already delicious before you start. Koji works in concert with practically any food preparation technique you know, with very little adjustment.

On the sweet side of things, we are forever chasing ultimate fruit ripeness for the sugars. Sweetness is another flavor element that is an indicator of essential nutrition, and one that we are always craving. Picking and eating berries at the peak of flavor in the summer is the best. However, it's uncommon to constantly have access to fruit at its optimal ripeness. As a result, we make adjustments by adding a touch of sweetness, sprinkling a touch of sugar or drizzling a little honey to bring balance. What if there was a seasoning that would allow you to sweeten the fruit with its inherent starch? If it's not quite ripe, a touch of koji will soften and sweeten it up to be just right. If fruits happen to be on the way out, a purée with koji will yield a sweet porridge that can be spun into a vegan ice cream. Sugar is enjoyable in itself—but the importance of kojifying food goes way beyond that. We'll show you how it can kick-start all the delicious fermentation processes we know and love.

## Why Isn't Koji Already Well Known?

Regional differences in our grain and starch staples is one of the key reasons why koji has not spread around the globe by now. Wheat, corn, and rice are the three primary cereals that have powered human life since the beginning of agriculture. Once these nutritional bases were established in their respective civilizations, no one felt any desire or need to change. Also, consider the challenges of cross-country transportation prior to steam and internal combustion engines. At that time, the best form of long-distance travel was by boat; introducing new

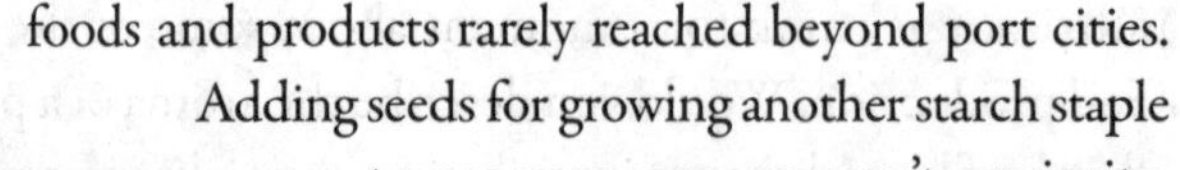

*Aspergillus oryzae* conidiophore. *Illustration by Max Hull, based on a Scott Chimileski microscopy image.*

foods and products rarely reached beyond port cities. Adding seeds for growing another starch staple to overseas cargoes wasn't a priority. In addition Japanese trade, and therefore exposure to delicious koji products, was closed off for two centuries prior to Commodore Perry's visit in 1853.

*Aspergillus oryzae* is quite unique when it comes to being harnessed for fermentation. It's not as simple as other wild ferments, where you can easily capture microbes from your surroundings. Only a few very specific isolated species out of hundreds in the genus *Aspergillus* can be used for koji; most of the rest are generally categorized as harmful to humans. It also requires very specific conditions on cooked whole grains in order to grow efficiently. That koji was originally discovered growing on cooked rice and not on corn or wheat makes sense, as the latter would have been ground and therefore difficult for the mold to grow on. However, the sprouting of wheat and corn did eventually lead to the discovery of malting to yield sugars to power alcohol production, a necessity to provide potable water. On the other side of the world with rice, koji provided the starch-to-sugar conversion necessary for making rice alcohol, toward the same end. This further reinforced the establishment of staple grain preferences that still exist around the world.

## Koji Is Bewitching

What is it that makes koji so attractive to a chef or other culinarian? For starters, koji is bewitching: Its life cycle is fascinating, and its aroma is intoxicating. Think for a moment about some of the other molds used in food production. Would you describe the aroma of a piece of charcuterie or a hunk of blue cheese as intoxicating, mimicking honeysuckle and tropical fruits? We think not. Molds used to make cured meats and cheeses often reek of such dank aromas as wet dog or basement, or carry the pungency of hard-cooked eggs or even chemical cleaners. Just based on this comparison, it is easy to see why so many people become captivated by koji—and we haven't even gotten to describing the seemingly magical qualities that koji imparts to nearly every food it touches.

Koji's alchemical abilities come from the many different enzymes that it produces. *Enzymes* are defined as substances produced by living organisms that act as catalysts to bring about specific biochemical reactions. The following are the main and most important enzymes you should become familiar with as you read this book and then venture off to begin working with koji.

**Amylase.** This enzyme breaks down the complex polysaccharide amylose into simple sugars such as maltose, glucose, and oligo sugar. There are many types of amylase such as glucoamylase and alpha-amylase. Koji also produces a lot of other enzymes that break down sugars called saccharases. While amylase may play a large role in fermentation, there are also polysaccharide-degrading enzymes that can produce sweet tastes beyond the basic "sugar." These sugars taste sweet on your tongue and are easily fermented with the aid of yeasts into alcohol, which can in turn be fermented into *acetic acid* (vinegar).

**Protease.** This enzyme breaks proteins into the amino acids that make them up. The most prevalent of the amino acids created is glutamic acid, especially when the food source is rich in proteins containing the amino acid glutamine. Glutamic acid and its derivative MSG are incredibly delicious and the backbone of the "fifth" taste (in addition to sweet, sour, salty, and bitter) known as *umami*. Umami provides a deep, rich, unctuous flavor to foods that leaves us satiated and fulfilled as we eat.

**Lipase.** This enzyme breaks fats into the fatty acids, esters, and alcohols that make them up. These are instrumental in the creation of the highly volatile aromatic compounds that give our foods the fantastic array of aromas that we encounter when eating.

Take a moment and think of a plainly roasted chicken thigh. Its dark meat and golden crispy skin are pretty delicious. Now imagine that you've used koji to unlock a tsunami of intensely flavored amino acids and sugars along with a plethora of fantastically pleasing aromas from that chicken thigh. It transforms from pretty delicious to the best-tasting piece of chicken you have ever eaten. Its flavor is so deep with umami that it feels as if it has become a permanent addition to your tongue. The aroma has transformed from simple roast chicken to one reminiscent of Parmesan cheese, toasted yeast, and aged meats. This transformative alchemy is what makes koji so enthralling. By simply marinating the chicken in a koji-derived ingredient such as amazake or shio koji (we'll go into great depth about these later on), you've transformed something from mundane to memorable. Your whole understanding of what *delicious* means has now been upended.

## Koji Is Universal

Functionally, koji is not tied to Japanese or any other Asian cuisine that uses it. It doesn't matter what the base ingredient is. It only happens that the microbe with this functionality, *Aspergillus oryzae*, was domesticated by humans for preserving food in Asia thousands of years ago. It does not taste like soy sauce, miso, or sake. It tastes like the flavor components of whatever you apply it to. Granted, there can be a hint of the distinct character of koji on the back end of the flavor in specific circumstances. When you think about the range of flavors found in a mold rind cheese, you can start to get an idea what we are referring to.

Practically every region in the world has a method of preservation that makes food more delicious as well as nutritious. No matter who you are, there's at least one preserved food, fermented or not, that you love. This was born out of necessity. For context, think about the last time the power went out for a few days and you had to figure out what to do with all the food in your refrigerator and freezer. You cooked what you could, held on to what you knew would be fine, and probably ended up tossing the rest. Likely some of things that you knew would be fine were preserves of some sort. Hmm . . . Now consider a time not that long ago when there was no refrigeration and people had to preserve large volumes of food seasonally to sustain life. It was a completely different food landscape back then. Preserves—specifically ferments—were required for survival; no one had the luxury of throwing anything away.

Then consider how long humans have been on the Earth. It was not that long ago that we were hunter-gatherers without agriculture. Preservation of found foods was essential. Simply drying and salting meat to keep it longer eventually led to the product becoming more delicious. This happened through naturally occurring microbes that brought enzymes along for the ride to break proteins down into amino acids. Nowadays people inoculate charcuterie with very specific strains of microbes to yield a desired flavor that's consistent and well known.

So, you might be asking, how does this tie into the whole koji thing? Any process that develops flavor takes time. One key factor that drives the duration is the enzyme loading. Whether it be naturally occurring or specifically selected for a food application, every common microbe used to generate amino acids pales in comparison with *Aspergillus oryzae*. Koji is the powerhouse of enzymes. Aside from traditional methods of applying a koji marinade to meat and *katsuobushi*, an aged smoked fish, we have discovered that you can accelerate the aging of animal-based proteins. Applying koji to charcuterie allows the drying process to yield the standard water loss (or activity) in as little as one-third of the traditional time. When you consider the standard one-year minimum for a prosciutto, it really makes you think. You can also develop aged cheese flavors in two months instead of having to wait an entire year.

## The Authors' Flavor Adventure Together

Both of us have loved food since birth and developed adventurous appetites as a result of the wonderful home cooking we enjoyed in our families. These experiences throughout our formative years made us understand the importance of nourishing the people you love. The faces smiling at gatherings as a result of eating well and being taken care of are the definition of *soul food*. This is the core of our drive to bring a community together to share and support one another. At the forefront of this movement is spreading the magic of koji. Each of us has a strong fundamental understanding of what's possible as a result of years of practice and experimentation, but we come at it from completely different perspectives. Jeremy, a fearless professional chef who runs a delicatessen, is constantly looking at the practicality of serving quality and innovative food. Rich, an adventurous cook, is forever interchanging techniques and ingredients to find tasty combinations that simultaneously respect and challenge tradition.

Whenever we come across a new ingredient, technique, or process, we start by getting to know what's possible based on what's been done before. At the beginning of our adventure, countless traditional makes all over Asia had our heads spinning, as did glimpses into applications in the fine dining scene. However, the more we began to experiment with koji, the more we saw the unlimited potential of how koji can be applied to any food preparation. To this day, both of us are still deep down the rabbit hole of experimentation and haven't yet found any sign of an end to our flavor journey. Koji is an unparalleled seasoning that can be used as simply or complexly as the cook who wields it, no matter the experience level. It's a true secret sauce—that won't be so secret after you read this book.

Before we knew each other, when we each independently realized how amazing koji was, the first thing we did was start to share with the culinary community. We wanted to know who else was out there diving deep; who could we share our experiences and ideas with; what more could we learn by connecting with the vast network of chefs, cooks, farmers, scientists, journalists, historians, craftspeople, and artists from all over the world. How could we all help one another grow in koji?

## About This Book

*Koji Alchemy* is a comprehensive primer for anyone who wants to understand the fundamentals of koji: how to make it and how to use it in a whole host of applications, ranging from savory to sweet. Our intention is for you to use this book as the sounding board for your most imaginative gastronomic innovations. We want you to take all that is in these pages and expound upon it, transform it, turn it into your own. Our endeavors with koji started with our own investigations and led us here. We can only begin to imagine what fantastical workings and explorations you will discover with this book at your side.

The journey we are on is driven by our desire to encourage people to make better food for one another. We hope this book helps you to understand the fundamentals without feeling daunted. In fact, if you look at the core of the recipes for each koji application, it's pretty much just mixing things together and waiting. Granted, there can be a whole lot more . . . but there doesn't have to be. We want you to see the beauty in marrying ingredients with technique and process, without the boundaries of following strict rules. And more than just making tasty food, we hope this encourages you to build strong communities and support one another when you're stuck, struggling, suffering, or you need that glimpse to get to the next level. Because when we recognize we're all in this together, we are all better off.

As you make your way through this book, be mindful that what we present is what has worked for us and the wonderful contributors who have generously shared. We have gathered our understanding of how to use this magical ingredient based on ancient preservation practices and collective experiences. This is not to say that there aren't a multitude of other ways to work with koji. Part of our journey has been taking not only what we want to understand more about and investigating it further, but also finding ways that koji fits into each of our styles of cooking. In this spirit we want you to use this book as a launching pad for your own ideas. Take what you find inspiring and experiment with it, craft and transmute, and allow it to evolve to suit your needs. Create your own magnum opus. There is no right or wrong way to express yourself and your likes in the kitchen. If you find something to be delicious and it brings you joy, then by all means embrace it.

In the following chapters we'll be discussing the many ways you can use koji in your kitchen—from amino pastes and sauces to an age accelerant for meats. Nearly everything you could want to know about the current applied uses of koji can be found. One thing to note is that our metric conversions are approximate and not exact—we often round the conversions to the closest whole number. With this book and a basic understanding of cooking, you will immediately be able to make your food tastier by incorporating koji. Okay, let's make some magic.

Welcome to the passionate obsession.

## KOJI ALCHEMY HAS ARRIVED

### Coral Lee

*As a result of all the amazing connections we have made in the koji and fermentation community, we are sure that koji is not just a passing craze but in fact something much bigger. Coral Lee, our food-minded friend, gets it. Here's her introduction to who we are and where this is going.*

*Koji Alchemy* has arrived, as if on cue. We now feast on keto jerky and chug adaptogen sodas in the hope of facing the drudgery of life as our best possible selves. We've grown to expect maximal flavor with minimal effort, willingly sacrificing what's true, natural, authentic for the sake of the hyper.

We're also all getting tired of being made to feel bad for making food that will never be as good or as true as our grandmothers'; and of overpaying for "fancy" food that leaves us wanting more (heart? soul? . . . satiation?). We all know that the future of our food needs to be equitable, accessible, and sustainable, but aren't totally sold on eating bugs or seaweed.

Koji is proof that good food isn't exclusive to the French-cottage-chic kitchens of culinary Luddites, nor the confines of an Instant Pot. There is no romantic stirring over a Dutch oven for hours; a $19.99 bag of sushi rice from your grocer's "ethnic line" not required. In existence before the delineation of geopolitical borders, koji grows on whatever you have, wherever you are, whoever you are. It simultaneously magnifies and celebrates familiar aromas alongside funk. And with the volume turned up across the board, Rich and Jeremy change how we feel about funk, mold growing on our food, "off-cuts" (hint: there are none), and free glutamates.

I was introduced to Jeremy at a time in my life when I was especially concerned with the hard-and-fast boundaries of (in) authenticity. He was the guest on the pilot episode of my podcast, *Meant to Be Eaten*, a show on Heritage Radio Network that explores cross-cultural exchange as afforded by food (or, why we eat, how we eat, and what we eat). When Jeremy explained that his eatery Larder Delicatessen & Bakery had not only had all of Japan's, but even its koji mascot's, stamp of approval, I took a step back. I believed, so much, in the just wielding of the word *authenticity* (and even more so, *inauthenticity*). I conflated authenticity with goodness, truth, and honesty. I wanted retribution for all those tossed lunches. Growing up alongside Jeremy and Larder these past two years, I've come to understand the conversation is not black and white but, really, infinite shades of gray.

I first heard about JD, or Jean Dough, or the brains and brawn behind OurCookQuest, while deep in the /koji Reddit-hole. I was looking at extreme close-ups of meat and sinewy squash innards, hoping for answers. "I know a guy," a user pm'd me. And so began my frantic pen-pal-ship with Rich.

Rich came armed to our first in-person meeting with at least twenty vac-packs. There was the usual spread—cheese and crackers—but also *gochujang umeboshi* and a Coca-Cola, and then the former plopped into the latter. What was supposed to be an interview—about him, his work, his belief that all anyone curious needs is the method—ended up being an hour of me tasting things. Before we parted, he handed me his last apple *hoshigaki*—so green-apple-tart it tasted like the color green. "How long did this take you?" I asked, cradling the baggie in both hands, unbelieving that anyone would part with such an involved project. Rich looked at me like I was crazy and shoved it back my way. Cooking in kitchens has taught me that those who hustle harder, guard their hearts, and only care about themselves and their wants are rewarded. The air-dried fruit was a physical reminder that there still are good people out there: who have studied and respect the past, but also are deeply anxious about how we move on from here. And who don't just want to gouge you.

Jeremy and Rich prove, with koji, that studying and understanding the flavorscape native to our foods is still relevant and revelatory, that there is still much magic to unpack and tinkering to be done. That you can make food instantly more tasty but also truthfully. When working with koji, wherever your ancestral roots with food lie is beside the point. What's more "authentic" to an ingredient than allowing it to taste more like, well, itself? We are all so very touchy and nostalgic when it comes to foods and memory. But nostalgia for that elusive something that needs to be preserved—as is, just so—can become dangerous, limiting, a hardening of and retreating within our shells.

Jeremy and Rich are lifelong learners and proof that earnestness is not dead, and neither is generosity. They are critical without being cynical. They expect the same curiosity and heart, if not more, in their co-travelers; and they foster both by encouraging experimentation along with failure. That is how #KojiBuildsCommunity. Jeremy and Rich foresee a future of food that is equitable, accessible, environmentally sustainable, and fun and delicious.

Fad foods remain as such because of cost, ineffectiveness, or both. What makes koji more than just another umami-bomb fad? Its application across cultures, ingredients, and methods. Pickles, naturally leavened breads, cured meats, and alcoholic beverages to name a few—all complicated processes relegated to "the pros," aka machines, are reclaimed for the home by the home cook made confident and empowered by koji. Pre-koji, my experiences with charcuterie (consuming, definitely not creating) had felt so fleeting, so wispy, so salty. With the help of koji, I slapped some dust onto a pork loin, and three weeks later not only tried *lomo* for the first time but shared that perfectly malty meltiness with over thirty friends on my holiday gift list (and still have chunks in my freezer saved for emergency *amatriciana*). To understand and properly leverage koji—its processes and flavors—is not at all difficult to do. Koji does not foster dependence or persistent, very good rule following but teaches home cooks to observe for themselves, reflect, and break some things some more.

CHAPTER 1

# What Is Koji?

You're likely reading this book for one of two related reasons. Either you don't know what koji is and you want to learn more about it, or you do know what it is and want to learn more about it. We don't blame you, as we are learning more about it every day. Indeed, koji is an amazingly transformative and seemingly magical ingredient that has bewitched many people over thousands of years. As you'll discover in this book, one of the many things koji does is turn complex carbohydrates into simple sugars via powerful enzymes it produces in order to feed itself. It is a type of mold used in the production of many foods such as miso, soy sauce, sake, *jiang*, *douchi*, amazake, makgeolli, meju, and *tapai* just to name some. Koji has been used for millennia throughout Asia and most recently, in the past 150 years or so, has been slowly conquering the rest of the world in ways that the people who first domesticated it could hardly conceive. Take the charcuterie that Jeremy makes at Larder Delicatessen & Bakery. After the meat is cured and inoculated with koji, the drying time is cut by as much as 60 percent. Imagine being able to make a prosciutto in six months instead of two years.

Koji is an extremely powerful organic technology that has not only shaped the foods of various peoples but also ingrained and transformed their very cultures. Actually, virtually every culture that encounters koji or a food made with or from it becomes entranced by its transformative power. The Japanese have declared it their National Mold and have even created comic books in which it is featured as a cartoon character! We feel that in order to truly understand what it is and what it's capable of, we must know a little about how, where, when, and why it came to be. When investigating these matters, it's always best to start at the beginning in order to give a complete understanding. That beginning would be koji's evolution.

## The Origins of *Aspergillus oryzae* (aka Koji)

The koji mold, *Aspergillus oryzae* (or, as we will refer to it going forward, simply koji), has a bit of mystique surrounding its origins. Due to rigorous scientific research, we know that koji evolved when it was domesticated from the highly toxic *A. flavus*. Dr. John Gibbons at Clark University in Worcester, Massachusetts, is currently leading the way in research related to how koji evolved away from its toxic ancestor, by identifying the traits and genetic changes that accompanied the domestication of *A. oryzae*. To address this topic he and his team sequence and compare the genomes of *A. oryzae* and its toxic progenitor species, *A. flavus*. They then use computational genomics, evolutionary biology, and population genetics to pinpoint genetic differences between domesticated and wild genomes. When the genetic differences are present in genes whose functions are known, they design laboratory experiments to test how these genetic differences change the characteristics of *A. oryzae*. They're essentially attempting to use biology to understand the traits that ancient artisans selected for when they domesticated *A. oryzae*. These findings have evolutionary, cultural, historical, and applied significance, which makes this system so exciting to many of us.

Why would an individual decide to conduct research on something so specific as koji? Well, when Dr. Gibbons went to college, he wasn't fully aware of filamentous molds and their uses. He had always been a big-picture person and applied to graduate school knowing that he wanted to study genomics, indifferent as to what his specific organism would be. After interviewing with labs that studied fruit flies, humans, yeast, plants, and molds, Gibbons ended up joining Dr. Antonis Rokas's lab at Vanderbilt University, where one of the major research areas was *Aspergillus* (the koji genus) genomics. Gibbons was given a stack of papers to read when he started at the Rokas Lab so that he could see which areas would be of most interest for him to research. One paper he read completely blew him away. He had assumed that only plants and animals had been domesticated by humans, but a number of papers called *A. oryzae* "domesticated."[1] One particular paper detailed how the publishing group used a combination of chemistry and archaeology to determine the contents of a nine-thousand-year-old pottery jar from China. They were able to show that the pottery held a fermented drink composed of rice, honey, and fruit, and that this type of fermentation required a mold that was really good at breaking down starches into sugar. It confirmed that humans have been making rice-based alcohol for nine thousand years through the help of filamentous fungi! The biology and genomics, cultural aspects, and applied side of this topic were exciting for Gibbons.

With the promise of potential research ahead, Gibbons dove in on a quest to unlock the how, why, when, and where of koji's domestication. For more than ten thousand years, humans have been taming plants and animals for particular characteristics. For example, domesticated plants usually produce more fruits or seeds than their wild progenitors thanks to selective breeding. Domestication has a profound impact on the genome of any given organism, and koji is no exception. Specific mutations underlie many of the traits selected for in domestication. For example, a single change in the genetic code of maize (corn) from its progenitor teosinte led to "naked grains" as opposed to the nearly impenetrable kernels of teosinte that are encased with silica and lignin. These mutations were shaped by selective breeding over long periods of time; when you simply compare the *phenotype* (essentially the way something looks) of teosinte with that of maize, you can easily observe these changes. And while plants and animals primarily shape our collective knowledge of the genomic and phenotypic effects of domestication, a number of bacteria, yeasts, and molds were also domesticated.

We've read some of Dr. Gibbons's papers about koji and its pathway to domestication, and they are nothing short of fascinating! To understand koji's impact on various cuisines and cultures, we felt it important to establish an understandable timeline for its evolution and domestication. Dr. Gibbons points out that we define *domestication* as "the genetic modification of a species by breeding it in isolation from its ancestral population in an effort to enhance its utility to humans." As we've noted, the domestication of koji occurred at least nine thousand years ago. Noncoincidentally, according to Dr. Gibbons, this is roughly the same time rice was domesticated (the two often go hand in hand). It makes sense that as rice was domesticated, the mold that eats it, koji, would follow suit in its new agricultural home. He also points out that people have been selling koji dating back to the thirteenth through the fifteenth centuries in China, a considerable amount of time before Western science even knew what microbes were.[2] To put this into relative perspective, it would still be roughly three hundred years before Robert Hooke would first observe dead plant cells under a microscope, followed by others observing living organisms.

The exact answers to the questions surrounding koji's evolution and domestication are yet to be found, and the process is quite puzzling due to the fact that koji and its toxic ancestor, *Aspergillus flavus*, share 99.5 percent of their genome. But the important point is that koji has been used in food production in China since at least 7000 BCE, making it one of the oldest domesticated foods on the planet.

## Koji and Clean Water

As with the domestication of many grains and fruits, koji can trace part of its pathway to domestication to people's need for safe drinking water. Humans have been evolving for millions of years, and at some point during the course of our evolution, we lost the ability (one that many other animals retained) to avoid getting sick from ingesting contaminated water. Many pathogenic microbes exist in fresh water that can sicken us severely and in many instances lead to death. And one of the many things that sets us apart from other animals is the amount of waste we generate. From excrement to food scraps, agricultural runoff to industrial waste, we produce a large amount of trash wherever we settle. Even our prehistoric ancestors generated waste on a level that made water sources in their settlements dirty. As soon as humans settled an area, it was virtually guaranteed that there would be contamination issues with the water they needed to ingest daily.

While water treatment and purification technologies have ancient roots, it wasn't until fairly recently that we learned about how microbes behave and how they can make us sick. During an outbreak of cholera in 1854, for example, British scientist John Snow realized that areas sourcing from well water had a lower incidence of contamination compared with areas drawing their water straight from the River Thames, which was constantly being polluted with large amounts of raw sewage.[3] A few years later Louis Pasteur proposed what we now call *germ theory*, changing our understanding of not just pathogens but all microbes, leading us to find ways to combat the pathogens. Snow's observations on the cholera outbreak eventually led to the addition of chlorine to public water supplies in order to keep pathogenic microbes at bay. But how is koji involved in water purification? The answer to that is simple. Booze.

For most of prehistoric and civilized history, humans have made drinking water safe by turning it into fermented beverages (otherwise known as alcohol). But the brews that were widely consumed historically were fairly different from the ones we have today, which are consumed for gastronomic, cultural, and religious reasons, and recreational pleasures, as opposed to necessity. These historic brews, such as wine, beer, cider, or mead, were much lower in their overall alcohol content than the ones we see today and were mainly brewed spirits. No higher than 1.5 percent ABV, these brews were consumed by people of all ages whenever thirst needed to be quenched. The alcohol present was enough, for the most part, to keep dangerous microbes at bay and prevent them from killing us. Author Michael Pollan addresses this head-on in his book *The Botany of Desire* when he traces the history of the seemingly mythical John Chapman (aka Johnny Appleseed). Pollan points out that during the westward expansion of America, its new inhabitants found that safe drinking water was rare. John Chapman sold apple trees to settlers to take with them so that they

## KOJI FOR ALCOHOL

### Stephen Lyman

*Stephen Lyman is one of a few Americans to apprentice at various Japanese breweries and distilleries, and he has also authored a comprehensive book,* The Complete Guide to Japanese Drinks, *on the alcoholic beverages of Japan. Stephen is a shochu ambassador who works with all aspects of these beverages from education to production. His knowledge of shochu and sake is, in our opinion, unparalleled.*

Koji may not have originated in Japan as an organism, but its use was arguably perfected by the Japanese. In fact, koji is used to make so many Japanese food and beverage staples that the common name for *Aspergillus oryzae* is the Japanese word *koji* with no independent translation into any other language. It's often mistranslated as "malt," but that's incorrect; malting is the process of tricking a grain into converting its stored starch into sugars so the seed can germinate and then killing it before it can sprout, whereas koji uses an entirely different process to saccharify stored complex carbohydrates. In Japan soy sauce, miso, mirin, rice vinegar, *kasuzuke* (a sake lees pickling method), *amazake* (a sweet non-alcoholic rice beverage), sake, shochu, and *awamori* are all made using koji during the production process. And this represents only the more common uses of the organism.

There are two working theories about how koji arrived in Japan, and the mystery will likely never be resolved. The prevailing theory in Japan (where the local origin myths suggest that the people were descendants of the gods rather than immigrants from China) is that ancient Japanese discovered that steamed rice left in a hot, humid place grew a fuzzy mold that made the rice sweeter than usual. The next step in that narrative is that if the mold was left in some water, it would turn into a weak rice wine. While that may be true, there is an alternative theory that makes a bit more sense when trading patterns are taken into account. The second theory is that the Japanese acquired *jiuqu* (Chinese "yeast balls") from Chinese traders sometime during the Asuka period (538–710 CE). Modern microbiological analyses of these yeast balls have revealed forty or more molds, bacteria, and yeasts all living together. These are most often used in other Asian alcohol traditions. For example, in the production of Chinese *baijiu*, some grains or other fermentable starches, water, and yeast balls are all mixed together and left to ferment. The resulting alcohol mash makes a wild, funky, and thoroughly unique distilled alcohol by the end of the process.

could press the apples into safe and shelf-stable cider, and thus have clean water to aid in their conquest.

As previously metioned, the earliest evidence of koji's use comes from archaeological discoveries from the early Neolithic village of Jiahu in Henan province in China from the seventh millennium BCE, in the form of clay pottery that had residual chemical evidence showing it contained alcohol made from rice, honey, and fruit.[4] We know this because only koji (and a few other molds that can equally be referred to as koji) can safely convert the starches in rice into fermentable sugars. Preserved liquids inside bronze vessels, whose lids had corroded around the edges and created an airtight seal, were found later in the record, dating back to the second millennium BCE. Upon laboratory and chemical analysis, it was discovered that this liquid was actually alcohol made from rice. This was a monumental discovery: In fact, one of the most efficient ways to convert the long-chain carbohydrates in rice into simple fermentable sugars is to use a saccharifying filamentous mold such as koji.

## What Is Mold?

Koji is a type of fungus known as a *mold*, one of the most culinarily important and ancient types of fungi. It's important to know what molds, and all other fungi for that matter, are and how they act, if you want a firm working knowledge of what is happening when you work with koji. Knowing about molds will also help you understand the potential applied uses for koji.

Molds are microscopic fungi that grow in multicellular filaments called *hyphae*. These hyphae grow over the surface of a mold's food source and secrete a host of different enzymes in a process called *extracellular digestion*, which predigests the substrate so that the hyphae can easily absorb the nutrients it needs to survive. Dr. Gibbons notes that digestive enzyme production is one of the characteristics humans selected for in *Aspergillus oryzae*. Some molds, such as koji's ancestor and close relative *A. flavus*, are not only highly toxic to us but also considered disastrous agricultural pests. *A. flavus* produces yellow-green spores and grows mostly on dead plant and animal tissue in soil. To the naked eye, it would be impossible

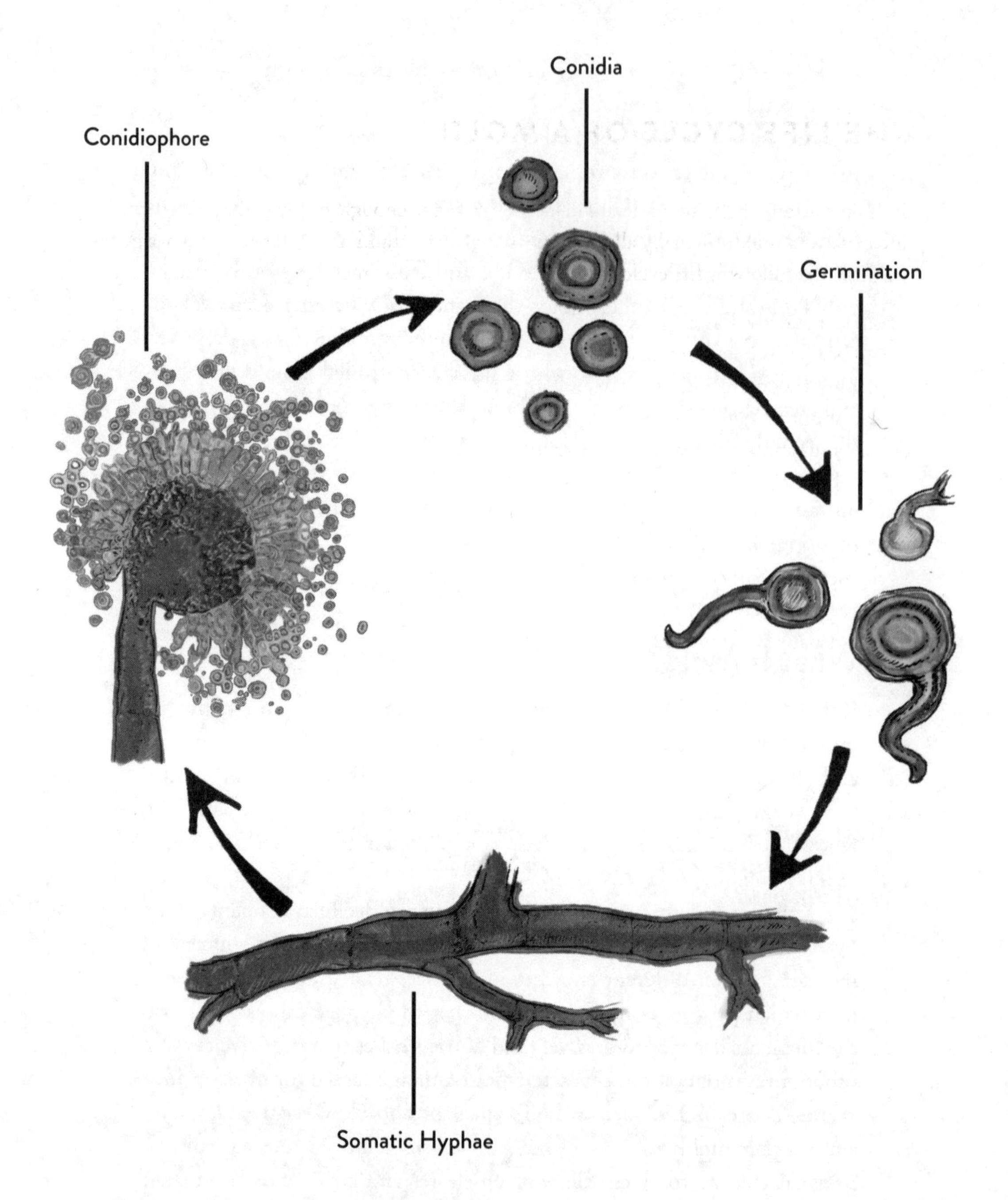


*Aspergillus oryzae* life cycle. *Illustration by Max Hull.*

## THE LIFE CYCLE OF A MOLD

It is important, first, to understand the life cycle of a mold. Virtually all molds go through the following life cycle:

1. First, it starts off as a spore. This spore can be loosely compared to the seed of a plant or the sperm and egg of an animal.
2. The spore then finds its way, typically via winds and air currents, to a substrate on which it can start to grow.
3. If conditions related to available moisture and temperature are favorable, the spore will grow into multicellular filaments called *hyphae*. These hyphae can visually be compared to the roots of a plant and functionally compared to our vascular and digestive systems.
4. As the hyphae grow over the substrate, they secrete powerful enzymes (which are types of proteins that we'll discuss later), which break the substrate down into nutrients that are easily absorbed into the hyphae.
5. Once enough nutrients have been consumed and the hyphae have formed a dense mat known as mycelium, the mold is ready to move into its reproductive phase.
6. During reproduction the mycelium produces *conidiophores*. These can be grossly compared to our reproductive organs, and somewhat resemble a lollipop or balloon on a string. The head of the conidiophore is, depending on the mold, covered with the next generation of spores.
7. Once the spores are mature, they are expelled from the mold via various mechanical processes and the whole life cycle starts over again.

As koji grows it produces visual, aromatic, and sensual cues that will let you know where in its life cycle it is. Knowing and interpreting these cues will allow you to fine-tune your use of koji for the purposes that we will discuss later in this book.

to differentiate between *A. flavus* and *A. oryzae* because there is so much phenotypic variation such as the color of their mycelium and the color of their spores. (DNA sequencing is probably the only way to know for sure.) Other molds such as *Penicillium nalgiovense* and *P. roqueforti* are indispensable in food production, especially for makers of charcuterie and blue cheeses. Understanding the differences in these molds not only is important for your safety in the kitchen but also will give you an understanding of what types of molds are best suited for a given culinary task and how to maximize the mold's effectiveness. (We will discuss this more in "Diving Deeper with Koji Spores" on page 99.)

It may be hard for some to believe, but it's important to understand that all fungi, molds included, are more closely related to us than to anything found in the plant kingdom. Therefore it should come as no surprise that koji (and its close relatives and other molds used in food production, such as *Aspergillus sojae*, used to make some types of soy sauce; *A. luchuensis*, used to make some styles of alcohol; and even *Rhizopus oryzae*, used to make tempeh) like and require similar environmental conditions to us humans. Moderately warm temperatures and levels of humidity coupled with indirect exposure to light and lots of oxygen are exactly what they're after and want you to provide for them. The great thing is that there is very little you need to actually do to cultivate this environment, as chances are it already exists in the very room you're in right now.

## Traditional Versus Modern Growing

In Japan, koji has been traditionally grown by sprinkling its spores onto rice cooked methodically by rinsing, soaking, and steaming until each grain can stand on its own, yet still be soft enough for the mold to penetrate. This rice is then placed into wooden trays made from Japanese cedar. The trays are placed inside an incubation room called a *kojimuro*, which translates to "koji chamber." The inoculated rice is carefully tended to around the clock to ensure that the best possible growth takes hold on the rice. This process is most notable in sake breweries, where controlling every possible variable is critical if you want to obtain a great beverage. The reason for this is that if undesirable aromas, tastes, or flavors develop, they will impart themselves to the sake. In the production of more complex foods, such as amino pastes, these undesirable features do not necessarily result in a bad product. Given the number of ingredients in an amino paste, the time it takes to age, and the fact that the final product will be combined with more ingredients and undergo other cooking processes, this attention to nuance isn't as important.

The wooden trays in which koji is grown are important and serve several purposes:

- They're durable, non-reactive, and have very little of their own taste or aroma especially when compared with other woods such as oak or apple.
- They're porous, which allows surfaces to maintain moisture without condensation and in turn retards pathogenic contamination.
- They're poor conductors of heat compared with other materials, which in turn makes them good insulators for keeping koji at the desired temperatures.
- They develop their own terroir in much the same way that a cheesemaker's milk buckets or butcher's charcuterie cave does.

Nowadays, in kitchens outside of Japan, growing koji in this manner isn't always feasible. The trays are expensive, and so is the skilled labor needed to attend to the koji. Due to this, many other techniques can be employed to grow koji. From the use of dehydrators and proofing cabinets to immersion circulators and metal sheet trays, we have discovered a plethora of ways to successfully cultivate koji in nearly any environment. In this book we'll explain and expand on these techniques. We will also continue to iterate that throughout history peoples with considerably less access to efficient technologies and a far lesser understanding of the sciences behind these processes have successfully cultivated koji. This will hopefully translate into you having the utmost confidence in your ability to grow the koji you want for your specific needs.

---

Koji's uses are very broad; the making of alcohol, as we previously discussed, is just the tip of the iceberg. As those who first used koji went through the trials and tribulations of producing alcohol, they began to observe many a happy accident. Chinese jiang, an amino paste similar to miso, appears in the *Wu Shi'er Bing Fang* around 200 BCE.[5] These amino pastes were one of the first major uses of koji after its original intended use, to saccharify the starches in rice. As amino pastes solidified themselves into the gastronomic fabric that enrobed most of Eastern Asia, other uses were developed. From the making of foods such as *bonito* (dried, fermented fish flakes) to various styles of amino sauces (such as soy sauce), koji has been indispensable in the culinary identity and traditions all over Asia.

Due to various factors such as economics, isolationism, and a general misunderstanding about koji thanks to losses in translation, koji remained a massive mystery to most of the world for millennia. But it still may seem puzzling that koji stayed exclusively in Asia for so long given the history of the spice and fabric trades that connected Asia with Europe. Dr. Gibbons speculates that this had more to do with the lack of availability of fermented food substrates (the foods that koji grows on, namely rice and soybeans) outside of Asia. Koji is mostly used to ferment soy and rice, but these commodities were introduced to Europe within only the last eight hundred years or so. If rice and soy were expensive for them to purchase, Dr. Gibbons imagines Europeans wouldn't have enough surplus to concern themselves with preserving them through fermentation. It seems that it was much easier for everyone involved to simply trade the foods and products made from koji rather than trade the ingredients needed to cultivate it and then produce these foods. These preserved foods—various amino pastes, amino sauces, alcohols, and vinegars—could withstand long-haul travel without refrigeration. Fresh-grown koji needs to be either used right away or stored in a manner that will preserve it while maintaining its enzymes in peak condition, and this was not efficient or easy to accomplish at the time.

## The Many Applications of Koji

With the advent of the internet and the current access that we all have to not just information but materials from all over the world, koji has just begun to be embraced on a world stage. You can now find koji and foods made from it in places far removed from its Asian roots. Koji is found in restaurant kitchens from Jeremy's casual and laid-back Larder Delicatessen & Bakery in Cleveland, to Rene Redzepi's awe-inspiring and captivating temple of fine dining, Noma, in Copenhagen, and every other type of restaurant in between. Koji is now widely available, and chefs from a variety of gastronomic identities are beginning to work with it in ways that harmonize with their native food traditions. In fact, koji is poised to transform the world's culinary cultures. From being used to make amino pastes from chocolate chip cookie dough, to using it as an inoculant for European-style charcuterie, koji's potential is materializing before our eyes. Larder is continually pushing to showcase the wonders of koji and has received many accolades for how they use it. Oatmeal cream pies are emboldened with an oat amino paste; pastrami is able to go from raw brisket to your sandwich in days compared with weeks; matzo ball soup is elevated to new delicious depths with the addition of a matzo ball amino paste; and vegetables are transformed to taste and eat like cured meat. All of this is possible due to the magic that koji brings to anything it touches. In our opinion the greatest thing about koji's ability to elevate and transform foods is that once you get the hang of working with it, you will see not only the ease with which you can use koji but also the never-ending possibilities.

But koji's applications are by no means limited to the kitchen. For many years now scientists, researchers, and inventors have been using koji to perform seemingly miraculous tasks outside of gastronomy. There are ways to introduce genes of interest that encode particular proteins into the *Aspergillus oryzae* genome and then take advantage of this species's ability to secrete large amounts of enzymes and other proteins. For example, *A. oryzae* was used in the late 1980s to produce a *lipase* (a fat-degrading enzyme) for laundry detergent, to break apart oil and fat stains on your dirty clothes.[6] There are also efforts to use *A. oryzae* in partnership with the yeast *Saccharomyces cerevisiae* to convert food waste into ethanol for energy purposes, and there's some work centered on identifying and exploiting *A. oryzae*'s plastic-degrading enzyme.[7] Something like this could potentially solve a great many of the self-inflicted environmental catastrophes that we have created. *A. oryzae* natural products are marketed as health supplements (for example, alpha-amylase), and there are a few studies suggesting the mold is a prebiotic in humans.[8] Alpha-amylase derived from *A. oryzae* is also an agricultural supplement for livestock. As you can see from these few examples, we are just beginning to unlock koji's full potential after roughly nine thousand years.

Dr. Gibbons believes that, in general, the microbiology of traditionally fermented foods is extremely understudied. Some traditional fermented foods have been around for thousands of years and hold tremendous cultural significance. For a food to be around that long, it probably has some beneficial qualities, and these qualities are likely in part due to microbial metabolism. This leads to many questions that will hopefully one day be answered: *What are the health benefits of fermented foods? What specific molecules confer these benefits? What microbes are responsible for producing the beneficial molecules? What genes underlie this microbial metabolic ability?* These questions are all at the root of our inquiries.

CHAPTER 2

# Creating a Common Koji Language

Chef Ferran Adrià is arguably the most brilliant and talented gastronome of the twenty-first century. He ran the three-Michelin-starred restaurant elBulli in Roses, Spain, and is considered the father of the Modernist Food Movement. Since closing elBulli, chef Adrià has been working via his Bullipedia Foundation on creating the world's most extensive lexicon of gastronomic knowledge. Inspired by this idea, we decided to organize a common language for koji. There are now people from all over the world using koji. Introducing a universal understanding of terminology eliminates some of the confusion about what individual regions have named various aspects of the koji growing and usage processes.

The current landscape of koji creations is widespread and somewhat confusing if you're just getting your feet wet. As with anything of significance working its way into the mainstream, a variety of terms are being thrown around based on each person's or people's source education and experience base. Most of the terms people use are rooted in fundamental Japanese techniques. This is in part due to the proliferation of Japanese culture around the world following the end of World War II.

The terms we will use and define in these pages are an amalgamation of current usage, the strongest influences from the most recognized Asian products, and a fundamental understanding of the driver of each different food made with koji. Our intent is not to re-create the already established food community vernacular but to provide clarity for universal communication. Without this there are intense amounts of confusion regarding koji and its intricacies. This is in no way meant to disregard the nomenclature used by other cultures to describe koji and its culinary makes. It simply serves to unify terminology for us newcomers. With further intensive study we encourage the use of culturally specific terms to specifically describe a food in its native context.

The language base is heavily influenced by Japanese terms, which is primarily due to the works authored by William Shurtleff and Akiko Aoyagi, who back in the 1970s researched and wrote extensively on everything under the sun about koji, the soybean, tempeh, and miso. The breadth and depth of their research is unparalleled to date and was way ahead of its time. (You can find all of their wonderful work at soyinfocenter.com.)

To reiterate the most fundamental term and definition, *koji* is the Japanese term for an *Aspergillus oryzae* inoculated grain or soybean medium to be used as a starter for transforming foodstuffs through enzymatic and fermented means.

**Amino Paste.** One of the best-known amino pastes is miso. Many know it for its namesake soup application. If you don't know much about miso, you can think of it as a close cousin to soy sauce but with a pasty nut butter consistency. The most basic miso consists of koji, cooked dry soybeans, and salt. There's a wide range of traditional varieties that differ primarily in the types and amounts of koji, base protein, salt content, and duration of fermentation. Both *autolysis* (the breakdown of organic substances by enzymes) and salt-tolerant microbes for fermentation power the flavor development. The key variables in miso are the ratios of the mix and the duration of storage prior to usage. There are two basic categories of miso: short- and long-term. At zero to three months of age, miso's flavor is primarily focused on the sweetness and fermented sugars from the high koji content. In general, anything held longer has more protein for umami and complex flavor development.

**Amino Sauce.** When it comes to popularity and widespread use of a koji product, soy sauce is the best-known amino sauce. At its core, the base ingredient of soy sauce is koji, made with a soybean-and-wheat mash that's added to a brine and allowed to ferment. Upon completion, the solids are strained out to yield the delicious liquid. In Japan it's known as *shoyu*, and in Korea, *ganjang*. As a logical outgrowth of this product, creative koji folks started making versions of their own with other proteins serving as the source of amino acids, which are the umami driver. *Amino sauce* is a broad term that can be used independent of ingredients. It also leaves the door open for any protein to be the flavor driver. We've used just about every ingredient we can imagine from insects to raw meat to make amino sauces, and so should you.

**Autolysis.** Autolysis is the enzymatic breakdown of an organic substance specifically powered by enzymes found within the cells of the substance itself. It is most commonly associated with decomposition as a whole and can be easily harnessed and controlled in the kitchen when we age meats. Koji and other fungi use a process called *extracellular digestion* to break down their food. Extracellular digestion is when an organism secretes enzymes to break down an organic substance into smaller nutrients in order to absorb them as food. These two terms were all that we could find in our research when investigating the how and why of koji's workings, but neither of them seems to perfectly match what happens once the mold itself has died or gone dormant, yet the enzymes it produced continued to work, as in the case of both amino sauces and amino pastes. So we use *autolysis* as the term to describe what is happening when we leverage koji's enzymes. This is important in order to differentiate enzymatically driven processes in these foods from the microbially driven ones carried out by single-celled beneficial bacteria and yeasts, otherwise known as *fermentation*.

**Amazake.** Amazake is a sweet, lightly fermented watery grain porridge that is likely the discovery that led to sake and *Aspergillus oryzae* in general.

It is simply a mix of koji, cooked grain, and water that exercises the amylase enzymes produced by the mold to break starches down into oligo sugar and glucose. This process is analogous to, yet dramatically different from and more efficient than, the malting process for beer making. As varied and complex as the flavors of different grains can be, the potential for amazake extends beyond that; the multitude of flavors that bloom as a result of autolysis and fermentation are amazing. Aside from being a sweet medium for all sorts of dessert and pastry applications, it contains a hint of umami derived from the inherent protein in the grain that brings the depth of flavor up just enough to make it more delicious than a simple sweet porridge. Lacto-fermented flavor also comes along for the ride, lending a level of funk and acidity to balance out the sweetness. Another wonderful thing about amazake is that it's loaded with protein enzymes that have not been used, which makes it a nice replacement for freshly made koji.

**Amakoji.** Amakoji can be thought of simply as a thick amazake focused on sweetness with very little fermentation.

**Kojizuke.** *Kojizuke* is the generic term for koji pickles. More specifically, *bettarazuke* is traditionally a koji-pickled daikon. The pickling medium is basically amazake with salt and a low-alcohol beverage, traditionally sake. Duration is a key factor in manipulating the flavor profile to be closer to either a sweet kraut or a more developed vinegar-based pickle. Any of these applications will have the umami backdrop that we have talked about throughout that makes it more than just a pickle.

**Shio Koji.** *Shio* is the Japanese word for "salt," and shio koji is the most powerful short-term protein marinade; it makes a piece of meat taste great overnight, or within a few hours. It does this by enzymatically breaking down the applied protein into amino acids, creating intense depth of flavor (umami). Making shio koji is as simple as mixing koji, water, and salt together; then you're good to go. Shio koji's high salt percentage limits fermentation, so that aspect has little to no flavor influence on the food.

Before we go any further, we'd like to shed light on an important aspect of discovering a food for the first time. When experimenting with new flavors, we must be willing to shed our preconceptions of what is "delicious."

## MY BLUE CHEESE IS YOUR STINKY TOFU

**Kirsten Shockey**

*Our friend, fellow fermenter, and author of* Fermented Vegetables, Fiery Ferments, *and* Miso, Tempeh, Natto and Other Tasty Ferments, *Kirsten Shockey, has traveled all over the world to experience the diversity of preservation. As a seasoned veteran in trying funky foods, Kirsten offers a perspective that's important to consider whenever you embark on a new flavor journey.*

The number of foods and drinks throughout the world that are made tasty (and in some cases, made edible) by fermentation is astounding—it is somewhere around a third of all the food eaten. Fermentation is fundamental in the evolution of the human diet, "discovered" as a way to transform foods for preservation and better nutrition—not to mention simply transforming a toxic substance into a non-toxic one. These discoveries were based on the foods that were available to a specific area—be they fish, meat, cereal grains, dairy, vegetables, fruit, legumes, or tubers. The fermentation of vegetables is pretty ubiquitous, as is fermentation in the processing of grains and leavening of bread. Oh, and everyone found a way to make alcoholic beverages by feeding yeasts sugar—from fruit such as wines and ciders, or from releasing the simple sugar bound up in starch with enzymes, which might be chewing corn such as done with *chicha* in the Americas, malting grains in the West for beer, or applying fungus (koji) to grain in Asia for sake and other alcoholic beverages.

Peoples of every continent have leaned heavily on legumes for the basis of their diets. Yet if we use broad strokes to look at legume fermentation, curiously, most is tilted toward Asia with its many variations on bean pastes, amino sauces, tempeh, and natto (and all its cousins). Looking across the globe there is a bit of legume fermentation in Africa, none historically in Europe, and very little in the indigenous legume cultures of the Americas. There are some theories about why; the most popular suggests it has to do with Asia having the soybean, which posed digestion and nutritional problems when eaten unfermented. Looking only a little farther west to India and Myanmar, we see that lentils, grams, and chickpeas are often fermented. Another theory is that this lack of legume fermentation in Europe or the Americas has to do with mold ferments (*Aspergillus oryzae* and *Rhizopus oligosporus* or *R. oryzae*) being Asian, but in the West culinary molds play a role in mold rind cheeses such as Brie, as well as in salami and blue cheese. These are different molds and different applications, but it wouldn't have been a radical shift to grow mold on grains or legumes for flavor and preservation, especially since humans have adopted foods along trade routes forever. So why didn't koji and bean fermentation make the journey? All this to say that in many ways it is a bit of a mystery why Europeans and Americans, despite hundreds of years of trade and exchange with Asia, with the exception of soy sauce, are still for the most part still unfamiliar with Asian ferments. Sure, many Americans go out to Asian restaurants, and we have wholeheartedly adopted raw fish (sushi) over the last three decades, but to integrate the fermented ingredients in our own day-to-day cooking has yet to happen. Despite Robert Rodale's 1977 prediction, "Before long tempeh will be eaten widely and lovingly across this land of ours," tempeh is still not found in every fridge in the US. Despite our love of marinades and scrumptious sauces, we still don't think to use gochujang or *doenjang* or even miso as that tasty something to add to our food. Even more surprising is that many of these foods are crazy convenient (miso soup: Just boil water and add a spoonful of miso!). And even though fast and easy is our jam in this country, most people still are unsure what miso is.

Here is the thing—our eating habits tend to follow the path laid down by our parents with those first bites of solid food and then are shaped by the culture around us. To complicate matters the microbes in our gut drive our cravings, and most of these microbes are the core population we inherited at birth from our mother (also out of our control). Which microbes thrive depends first on that early diet and sets the stage for what we crave. Eventually we make our own decisions, and you can almost visualize behind our outstretched hands reaching for what we buy or pull out of the fridge the history of the people, the lunchrooms, the advertising, the warm memories (and the painful ones) around food, making that choice with us. So as delicious as these unfamiliar foods are to a large portion of the world, they are not familiar to most Westerners.

On a chemical level many of the compounds that make a creamy ripened Camembert cheese a so-called stinky treat are the same that give natto its so-called aromatic deliciousness. This is in some ways a question of texture. For example, here in the West we are squeamish about natto's stringy texture, yet we love melty cheese, which often brings about the same feelings of disgust among Asians. When you start taking apart the flavors and the taste spaces that foods around the globe occupy, they are not that different.

Despite the fact that it sounds impossible to change our tastes and eating habits, individually or collectively, it is entirely doable. People are omnivores, programed to learn to eat the things we like, but we can learn to eat what we might initially dislike. (Think about bitterness—a kid might think coffee is gross, but what would we do without it?) It's time for us to take the journey, enjoy the taste adventure, and create new flavor traditions—legacies that will live in our cultures long past the experimenting we do in our kitchens.

This is of the utmost importance with fermented foods and especially foods made with molds such as koji, which are made with microbes that we've been indoctrinated to be fearful of.

## Koji and Cultural Appropriation

Food is fundamental for life, and, outside of water, humans require three basic components—proteins, carbohydrates, and fats—in order to survive. Long ago, people figured out how to sustain life with what was available in their respective areas, starting with understanding their land and surrounding resources. They then evolved methods of making nutrition more readily available through agriculture, methods of preparation, and preservation. Local and regional cuisines everywhere are based on what was accessible and the specific gastronomic innovations that arose within those given areas. In the past, they were limited by transportation and storage technologies. But in modern times we no longer have these restrictions; you can get pretty much any ingredient you want shipped to you the next day. There's something to be said for a pantry that has no bounds. So when we are asked whether making foods with koji constitutes cultural appropriation, it doesn't make sense.

Cultural appropriation is based on misunderstanding and lack of respect. This is the complete opposite of what we practice every day when we make food with koji. In terms of understanding koji, we have done

## KOJI: SOMETHING TO WRITE HOME ABOUT

### Cynthia Graber

*Cynthia Graber, an insightful food journalist and cohost of the podcast* Gastropod, *has a knack for seeking out some of the most intriguing stories in food. Once she caught on to exactly what we were up to with koji, she quickly figured out this magical mold is not a trend but something that will forever change the way we think about and make food. This is her koji story.*

A few years ago a mutual acquaintance suggested I meet up with Rich Shih, who was making all sorts of unusual varieties of miso at his home in Massachusetts. I was intrigued. I was already obsessed with miso—I often ate miso soup for breakfast—but I'd never seen it made. Rich was breaking with miso tradition, crafting the fermented paste out of not just traditional beans, but also ricotta, bacon, and peanut butter. But the tastes of those miso varieties, while delicious, weren't what left an indelible sensory mark on my memory. Instead, the most visceral memory from my visit to Rich's home-based food lab occurred when he lifted off the plastic cover of his jury-rigged insulated cooler to show me his koji incubator. I fell in love.

Koji, the fungus responsible for transforming the proteins in beans and other protein-rich substances into miso, was growing on damp, partially cooked rice in the warm, humid cooler, which had been tricked out to re-create the environment of the microbe's origins in the tropics of Asia. Koji enveloped the rice with a white, fuzzy, cloudlike blanket, and released a smell that was simply intoxicating, a bewitching mix of grapefruit and flowers and funk. I wanted to bathe in it. I would wear a koji perfume.

I could easily imagine how koji had captivated people who stumbled upon it growing on a damp batch of grains, thousands of years ago, probably as long ago as milk-loving cultures started refining the microbial life that gave us cheese. These early koji experimenters decided to keep their moldy, grapefruit-scented starches around instead of throwing them out. From there, somehow, they figured out that the fungus-grain mixture could then be used for a second fermentation, to create foods that are now staples throughout much of Asia: miso, soy sauce, and sake in Japan; Chinese fermented black bean paste and rice wine; Korean fermented bean paste.

I wrote about Rich's untraditional varieties of miso for the *Boston Globe*, but my koji obsession was far from over. "You have to talk to Jeremy Umansky," Rich told me. Jeremy is a chef in Cleveland who's been testing koji's limits as well, growing the mold not only to make miso but also to cure meat. And so, for a feature article for *Cook's Illustrated* (and later for an episode of the podcast *Gastropod*, which I cohost), I flew out to Cleveland. There Jeremy opened up a whole new world of cooking with koji. He marinated chicken in koji, and the meat caramelized beautifully in the pan, the flesh itself even more tender and richer than usual after its enzyme bath. And he did something that wasn't quite within health codes: He grew koji on rice flour directly on meat at those warm, humid temperatures in an incubator. This would usually lead to a rotten, gut-twisting meal. Instead, the pork and beef I tasted had cured under a thin, fuzzy microbial crust, koji crowding out the harmful microbes and creating a new world of meaty flavors. It was delicious and unlike anything I'd ever eaten.

I dove further down the koji rabbit hole. I spoke with microbiologist Joan Bennett, who teased out the history of Jokichi Takamine, a Japanese man who first brought koji to the US in the late 1800s; he tried to use it to transform the American whiskey industry, but instead developed a popular koji-based digestion aid that made him so rich that he helped finance the Japanese government's gift of cherry trees, ones that still grace the Tidal Basin in Washington, DC. I met with John Gibbons, a scientist in Worcester, Massachusetts, who's applying genetics to try to re-create koji domestication in his Clark University lab to understand its evolution.

And of course, I cooked with koji. I used shio koji, which is a liquid or paste made of koji-inoculated rice plus salty water. One time, I marinated thin fillets of ocean perch I'd picked up at the farmers market for a few hours. I rinsed the fillets off, patted them dry, then seared the fish and finished it off briefly in a hot oven. The result: slightly cured, savory ocean perch that resembled one of my favorite foods, miso black cod. It was mind blowing.

extensive research and are well read and educated when it comes to the traditional methods of usage. We hold the highest regard for those who have dedicated their lives to continuing to make amazing koji-based products that are unparalleled in quality. Of course we are very much inspired by and romantically influenced by traditional techniques, but the expanded ideas represented here in this book are the outgrowth of a potential that only recently has begun to be realized.

## Koji at Larder

Jeremy owns and runs Larder Delicatessen & Bakery, an Eastern European delicatessen with Jewish food roots, with his wife, Allie La Valle, and friend Kenny Scott. They've always faced the question of why anyone would use koji in the context of a Jewish deli. Aren't the traditionally prepared deli foods already wonderful on their own? Generations of Jews have been using the same recipes for years with success. Most cooking and baking recipes and techniques at ethnic restaurants are passed down by hands-on training from generation to generation. Even if there are documented recipes, chefs understand the fundamentals and take liberties based on multiple factors such as flavor preference, available ingredients, what customers like, budget, and so on. For sure, there's a fundamental method to prepare a particular dish that makes it what it is, but it's natural to be flexible and make improvements. It's what good cooks do.

The Larder family is all about feeding guests tasty food they'll enjoy. They all have culinary training along with thirty years of combined experience among them as professional chefs, so a fundamental understanding of how to optimize flavor is ingrained. Jeremy grew up watching generations of his family cook, inspired by Jewish food and learning how to make it. It only makes sense that they would use their collective knowledge to make the best Jewish deli food they can.

What food makes for a great delicatessen? Pastrami, of course. What makes for an awesome pastrami? Tenderness and juiciness, of course. Through traditional marinating and low and slow cooking, you get a delicious product. The marinade not only infuses the meat with flavor, but also offers some level of tenderization. This is followed by smoking, a low-heat cooking process conducted over a long period of time so the collagen in the connective tissue dissolves into the meat and makes it unctuous. Is that enough? Well, if you've ever smoked brisket before, you know well enough that some sections of the meat are stellar and others aren't. This is a result of the inherent inconsistency of a cut of meat. No matter how much love you put into a low and slow process, there's always an area that suffers. However, if you add koji to your marinade (as we will explain in the "Enzymes for Umami" section on page 46), you can leverage the enzymes to break proteins down, tenderizing the meat even more and increasing the level of umami so every bite that comes out of that cut is great. Sugars also come along for the ride, which allow for the complexity of caramelization, basically making a natural BBQ sauce with the meat juices.

For vegetarians and vegans, mushroom pastrami and root vegetable charcuterie are also on the menu at Larder! There's something to be said for a curing process that can create flavor in *anything*. Koji can take the touch of inherent protein in the base (vegetable) and koji itself to create enough amino acids to make it wonderful.

Cooking is all about making the best-tasting food with what you have to nourish and please whoever is sitting at your table. Whatever your dietary choices and the reasons behind them, the philosophy of what we are writing here will help make your food better. The core of this book is all about providing you with the basics to get there. This is a universal practice of cultures everywhere.

## Koji Is a Culture

What's wonderful about a bed of fresh koji is that it's a perfect habitat for all the tasty natural food microbe colonies we know and love. If you think about the process of making your own sourdough starter by mixing some flour and water together and leaving it out, koji as a fermentation culture makes all the sense in the world. It's exposed to microbes in the air throughout the entire time the *Aspergillus oryzae* is developing. Sugars produced toward the end of the process also help. With this fundamental understanding, we've been mixing fresh koji with a little water and letting it ferment at ambient temperature for three to five days. We call this our universal SCOBY, a symbiotic colony of bacteria and yeast. You probably know this acronym as the starter for a kombucha, but it can be used in a general sense for any natural fermentation culture. The koji culture can be used to begin or enhance any fermentation process. It can be used to

(*continued on page 42*)

## HONORING THE PROCESS AND LOCAL INGREDIENTS

**Sarah Conezio and Isaiah Billington of White Rose Miso**

*Some folks genuinely fall in love with koji and honor the process to bring us delicious makes. The insightful ones see that there's a growing desire for beautiful locally made preserves that are inspired by the traditions of old, and aren't afraid to build a business around something they believe in. Sarah and Isaiah are two friends of ours who represent all of this through their fermented goods companies, White Rose Miso and Keepwell Vinegar. Pastry chefs by trade, they understand how to make a range of flavors—from light and delicate to complex and deep—shine in a dish. They both possess a unique perspective that's tuned into the significance of quality for every ingredient that goes in. Who better to be making enjoyable ferments? When we thought about voices to represent the significance and impact of koji in America, they were among the first folks we turned to.*

We have been using miso in our cooking for about ten years. We were cooking in the types of restaurants called New American, united more by an ingredient sourcing approach and equipment than an actual repertoire of standard dishes or a cuisine. Since we couldn't, or had decided not to, range far afield for exotic ingredients to wow our guests, we relied often on firepower—no amount of char or caramelization was too much from our wood-burning ovens—and density of flavor and richness. More butter, or bright orange yolks, or powerful sauerkrauts and heavily aged beef replaced truffles and foie gras. We thought that humble ingredients such as our house-fermented hot sauce were magic, since we could amplify heat, salt, acid, and funk at the same time. That stuff was probably in too many dishes.

Miso paste was, then, a revelation. It packed a huge and concentrated punch, and was made of the humblest ingredients, the pantry staples that we were all used to trying to convince everyone were as special as passion fruit. It was exotic and tantalizingly local. We could rev up sweetness, umami, and salt in one go, and we could just fold it into anything at the drop of a hat. My first love, maybe predictably, was South River's Hearty Brown Rice Miso—we are from pastry, and that stuff is sticky—but we soon got lucky with a few decent early batches of our own. You can still find our love of those extreme flavors in our benne seed miso, for which the seeds are toasted almost to a char before grinding with unpolished barley koji.

Some fermentation processes are quick, passive, and idle. If you set kimchi up properly, all you have to do is taste it until it's ready. Lacto- and similar ferments should happen as close as possible to the mouths that are going to eat them. But as processes get longer, more complex, and more hungry for attention, they lose their ability to fit into a cook's schedule or the one closet where you hide all the fermentation experiments from the health department. We felt the draw of vinegar, miso, and shoyu, and knew that the slow accretion thereof could never fit, in any way meaningful to the growers it was our mission to support, in the high-capital-turnover square footage of a restaurant. We moved out into the country and started taking up lots of space.

Our mission is to interact as closely as possible with the totality of local agriculture. We support tomatoes in season, and berries in spring, sure, but fermentation allows us to work with ugly fruits, bulk buys during the gluts, and a wider variety of crops. If we want to eat something, so do bacteria. Koji is especially valuable to us in our range of activity because it allows us to support incredible grain farmers and bean growers while making shelf-stable pantry items that provide the same transparency of technique and provenance of inputs as do the chicken, beef, eggs, and produce of the chefs we work for.

Running a business based on koji is a mixed blessing right now. There's a growing awareness of koji and a fascination with its versatility. We have chefs asking us for fresh koji, shio koji, shoyu, shoyu koji, miso, rice vinegar, and more. We like to try to connect as many people as possible with what they want, and sometimes consistency in our availability is sacrificed for that variety.

That's really not the challenge, though, right? Just make more—that's what is difficult. Koji production resists easy multiplication. Of course, the cooking of grains and beans can be expanded, and a larger chamber can be bought or built to accommodate the increased volume. But we have found that obsessive control of temperature, humidity, time, and mold growth are very important to the eventual quality of the koji's final form—tiny flaws are compounded and magnified over the course of the long ensuing fermentations. You can build a bigger box, but that doesn't mean you can make good koji in it.

Koji is important to us because it makes everyday ingredients into something special, takes rice and beans and makes them more sweet, more earthy, deeper, and calmer, and then suspends that elevation in time. It's the ultimate democracy of flavor.

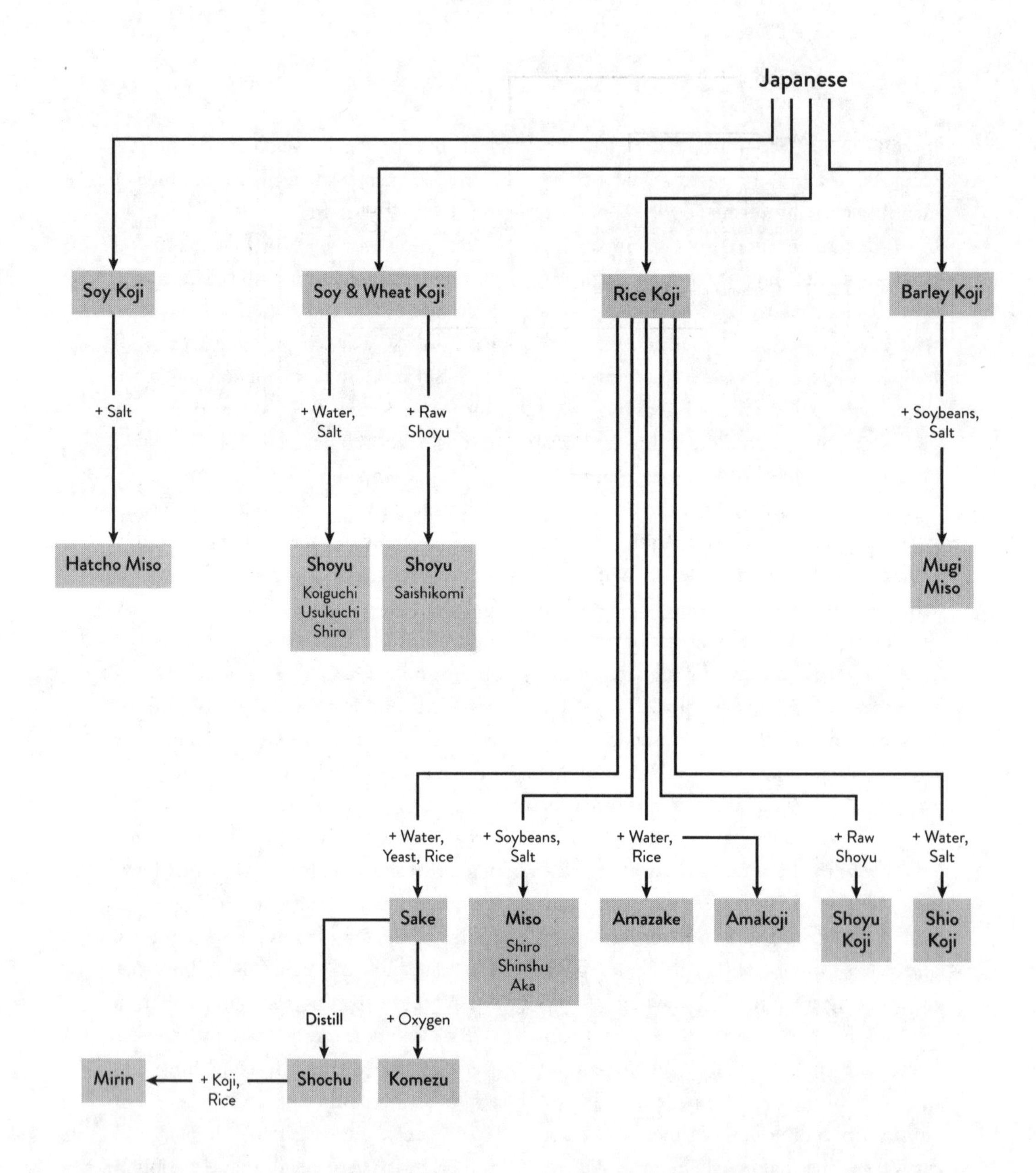


Map of traditional makes that use *Aspergillus oryzae*. In Japan, koji is primarily driven by this microbe. However, in China and Korea, it's only a component in their respective starters: qu, meju, and nuruk. This flow chart is the result of Sarah and Isaiah of White Rose Miso having the

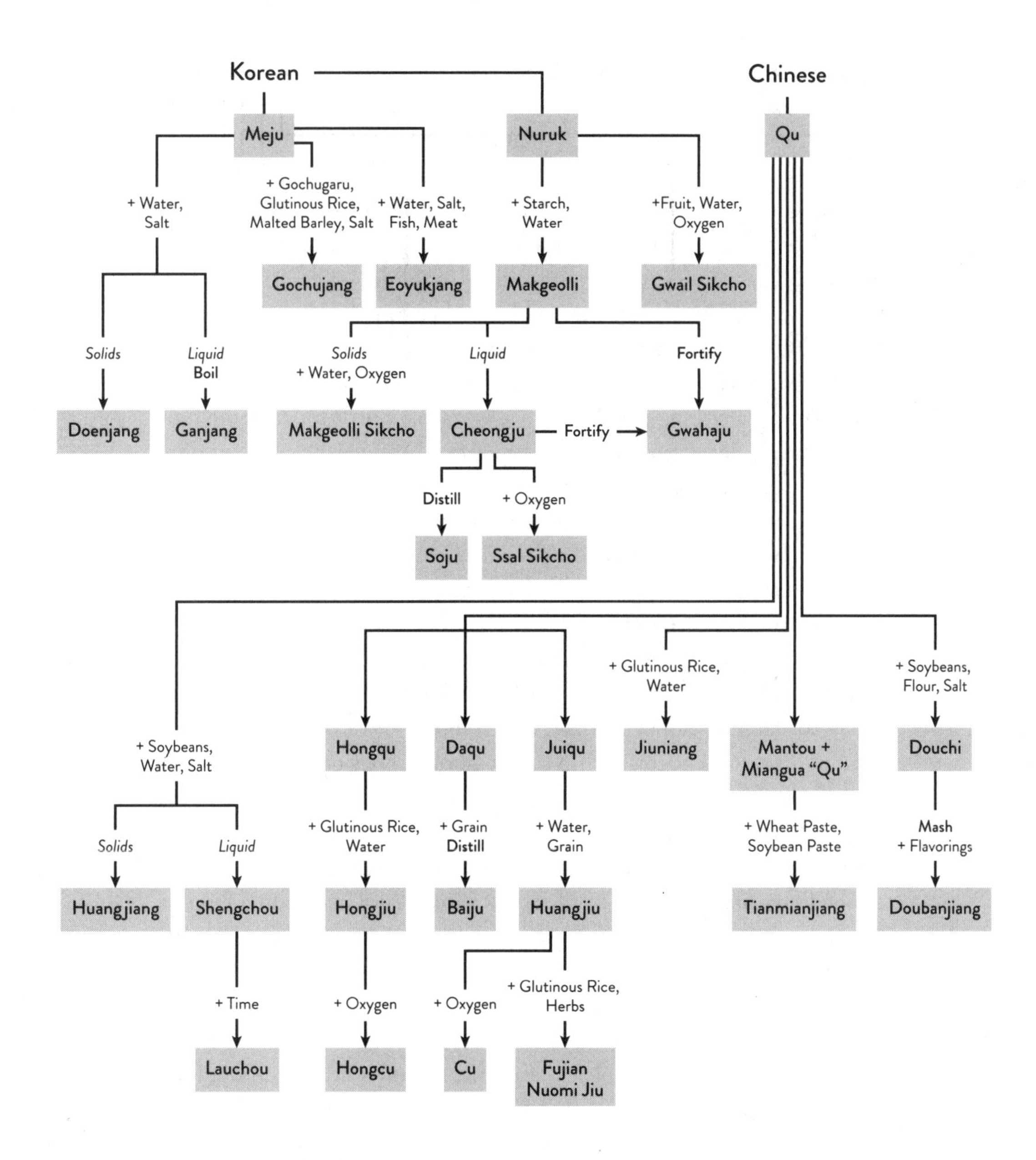


wonderful idea of showing the wide variety of what's been in existence for ages. Thanks to their work combined with contributions from knowledgeable friends, you can see how integral mold fermentation is in these countries.

culture cream, get a sourdough starter going, kick-start a hot sauce, be a vinegar mother, ferment cider, and more. The possibilities are endless and offer the benefit of adding a touch of umami that enhances flavor depth.

We mentioned earlier that koji is as universal as salt when it comes to cooking—one of the keys to understanding and unlocking the potential of koji. There are infinite possibilities when it comes to applying koji to any food preparation, but it requires a little bit of finesse to elevate flavors appropriately. What do we mean by that? Most people who start playing with koji are focused on maximizing umami. However, there's only so much deliciousness anyone can take before their palate is blown out and they can't taste anything else. As much as we take it a little too far from time to time, there's beauty in just a small amount. Consider the crunch of finishing salt on a chocolate chip cookie that elevates the flavor as you chew. The chocolate has more depth and the toasty caramel notes are enhanced as the tiny bits of salt dissolve. If a chocolate chip cookie can be improved by just a hint of salt, a single dimension of flavor, what happens when you add something crazy complex?

A related question would be, "Can you make a chocolate chip cookie better than any chocolate chip cookie you've ever had?" Well, we're not so bold as to claim we can make that happen for *everyone*, but simply adding a hint of miso to a chocolate chip cookie does a pretty good job. One key factor is the Maillard reaction, most commonly understood through the complex flavor that results from browning meat. Did you know that dark (long-term) miso also has flavors developed through Maillard? This is achieved over time instead of by the application of heat in cooking. Just one of many wonderful things that happen during the fermentation process and drive the unparalleled flavor of miso. Adding it to pastries and desserts is a must. We won't even talk about the cookie dough miso we made . . .

To quote Alex Talbot of *Ideas in Food*, "Make something more delicious than it is." This is a simple concept that anyone who cares about food is striving for, but it takes concerted effort to achieve. Just think about the difference between a vegetable stock with and without umami additions like mushrooms, tomato, or seaweed. All it takes is a touch to make something unforgettable. Once you start thinking about koji applications as an umami dial you can tune in, that's when your food starts to gain attention.

## 100 Percent Utilization

In today's world there are serious and grave concerns about our dependence on wasteful disposable goods. Koji gives people an outlet to keep food waste out of landfills and compost piles and in our kitchen. This book gives you the tools to use every last bit of any food you will ever encounter. Leveraging the enzymatic power of koji, you can coax flavors out of scraps of food that you wouldn't be able to use otherwise. Specific recipes for using carrot scraps for mirin and citrus peels (not just the zest, but all of the remains after juicing) for an amino paste are just two of the many interesting ideas we'll be sharing.

Within the pages of this book, there's a universal goal we are working toward. Simply put, we want people to help one another. The wave of koji interest is growing due to the community freely sharing ideas to make better food. We all want to understand more, but there's only so much each of us can do on our own. Everyone has limitations and needs help. Social media has provided an interconnectivity of people and accessibility to information that is relatively new to humankind. This can be a powerful thing! Over the years we've been sharing our knowledge and exchanging ideas with fellow fermenters and cooks from all over the world, which has allowed our depth of understanding to grow faster than it would have if we'd worked independently in our kitchens. We would not be where we are without all the wonderful people who support this culture that we've built around the love of food and passion for a better tomorrow. #KojiBuildsCommunity is where you can find all of us.

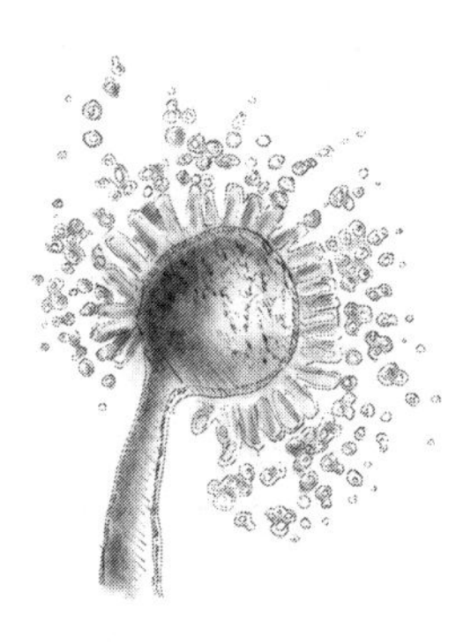

## CHAPTER 3

# The Flavor-Making Road Map

The rabbit hole of ideas and possibilities runs deep with koji. We recognize that it may seem daunting to novices on this flavor journey. The good thing is that all these processes can be broken down into very simple pieces that you can easily wrap your brain around. The key to the castle of koji is understanding how to work with enzymes. This process is not typically part of a chef's or cook's wheelhouse, but we feel it's essential to anyone who enjoys delicious food. Once you learn how to leverage the enzymatic actions that koji affords, you will be able to create an infinite amount of tasty foodstuffs to work with.

We have racked our brains in order to simplify the crazy web of interconnectivity between enzymatic breakdown and fermentation. How can we take all the concepts and condense them into a straightforward and easy-to-understand map for practical use? Unfortunately, we found that generating a chart to encompass all that koji can do requires more brain power, computer power, and time than we had when we started writing this book. (One day we hope to have this available in a virtual interactive space where it can live and grow with convivial input.) Fortunately, we were able to develop one basic master chart and a few primary-process road maps to help you see the forest for the trees. They're driven by to two things that we all love: umami and sugar.

## Enzymes for Umami

Let's start with the most important flavor element behind the foods mentioned in this book, umami. At this point, you are well aware that protease enzymes are the key to breaking down protein into amino acids to develop depth of flavor. The next step is to understand how to use them effectively. In order to do so, we have a chart that shows the interconnectivity of shio koji, amino sauces, and amino pastes. The relationships that we'll discuss are: enzyme access, salt, and water.

From an application standpoint, the links between an amino sauce and paste are clear, but shio koji is a bit different. However, if you think about them in the context of leveraging protease enzymes to create umami, the connections among all three starts to make sense. We can make it clearer by describing each process simply in the following two sentences. Shio koji is used to coat and infuse pieces of food with protease to create amino acids from the proteins it comes in contact with. Amino sauces and pastes maximize amino acid generation by optimizing protein-to-protease contact within the mixture.

Functionally, all three processes do the exact same thing, but by matters of degree based on the application. Shio koji is a short-term marinade that is intended to make a piece of food more delicious with its inherent proteins but not significantly change the structure or base flavor. Amino sauces and pastes are fermented and aged condiments that develop concentrated complex flavors and result in extremely broken-down protein structure. At the end of the day, shio koji is "umami to be" while amino sauces and pastes are already umami bombs.

Essentially, shio koji and amino pastes/sauces are two sides of the same coin. Why take the time to explain this? Well, it means that you are now empowered

Comparison of a surface coating (*top*) versus a mash/mixed medium (*bottom*) application. *Illustration by Max Hull.*

to use protease enzymes as an umami seasoning however you like and not be tied to a single limited process or koji-derived product. With this understanding, one application can easily become the other. The shio-koji-marinated fish you let sit too long in the fridge may now be mush, but you can repurpose it by making it into an amino sauce. On the flip side, you can put shio koji on cooked beans for a hit of umami that makes them shine in a salad. You get it now. Koji enzyme accessibility to proteins drives the umami level.

## Water

Water is simply a vehicle for getting ingredients to interact and meld with one another. It also dictates how the koji-derived food is utilized in application. If you think about the core of making amino sauces and pastes, the only difference is the concentration of water. For example, soy sauce, a well-known amino sauce, is created by infusing a liquid with the essence of the ingredients. By steeping a cooked soybean and cracked wheat koji mash in salted water, enzymatic action and fermentation are able to occur. Blending the mash is not that important. On the other hand miso, a fundamental amino paste, consists of cooked soybeans, koji, salt, and a bit of water all blended in order to facilitate the direct contact of all the ingredients to the enzymes and microbes needed. On the other side of the spectrum, shio

### UMAMI APPLICATION MAPS

To help you understand the interconnectivity of all the koji makes that drive umami, we've put together two important maps: surface application and umami condiments. Both maps list the drivers of each individual make so you can easily reference the similarities and differences between them.

The surface application map is just that. It shows all of the possibilities when applying koji to the surface of a piece of protein, whether it be a chicken breast, steak, pork chop, fish fillet, block of tofu, or a whole muscle cut.

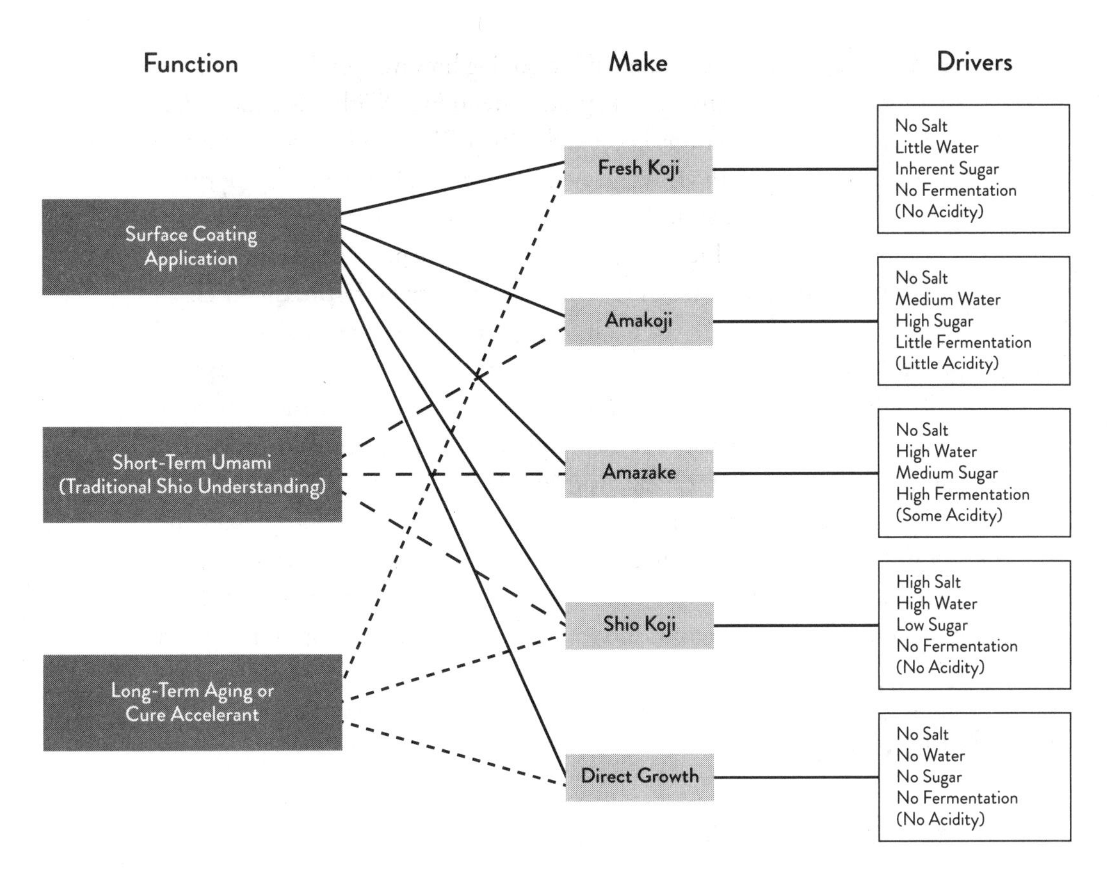


Koji surface application map: how functions are connected to key makes and drivers.

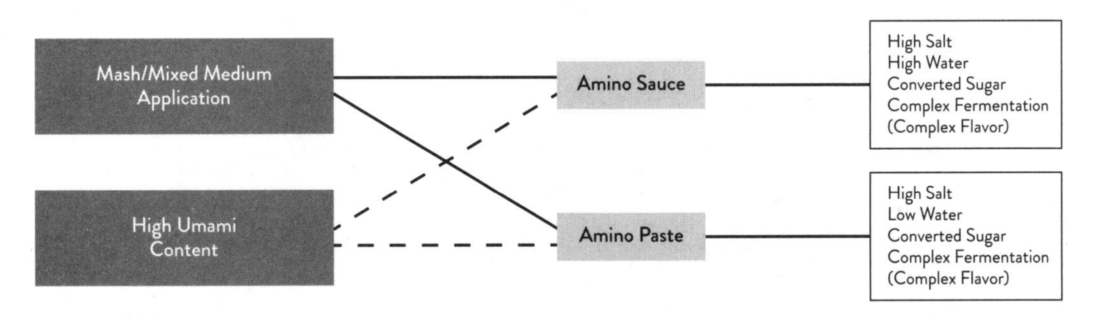


Umami condiment map: amino sauce versus paste comparison.

koji is water-based to aid the infusion of enzymes into whatever it comes in contact with. In any protein application, water is the facilitator used to maximize the creation of amino acids and infusion of flavor.

## Salt

Salt is primarily used for three reasons that have been the driver of all cuisine and survival since the beginning of cooking: seasoning, safety, and preservation. All three of the primary umami makes we mentioned previously are used in place of salt in food preparation because they add a lot more dimension. The salt concentration in each make is uninhabitable for pathogen microbes, but it happens to be favorable for salt-tolerant fermentation microbes that make amino pastes and sauces even more delicious. On the other side of the spectrum, shio koji is a super brine that infuses into whatever it's applied to. When you brine a piece of meat, you dissolve parts of the protein structure and create a condition that helps retain more water.[1] The bonus is the enzymes coming along for the ride to create tasty amino acids with the inherent proteins that make up the piece of meat. As with all food preparation, salt is essential to flavor and life.

## Any Starch for Sugar Feeds Fermentation

As we've previously discussed, koji contains amylase enzymes that break starches (amylose and amylopectin) down into sugars. Why is this important? It allows us to make sugar from nearly any starch to power all the major fermentation processes we know and love. Of course, all you have to do is grow *Aspergillus oryzae* on any starch or add supplemental starch then mix it all with water and wait. Also, any inherent protein yields a touch of umami that changes the game. Think about how delicious a horchata can be, a morning porridge, or oat and nut milks. This opens the door to so much flavor potential.

Let's talk a little more about amylase enzymes. Interestingly enough, everyone on the planet uses them to convert starches to sugar every day. They're in your saliva and in your stomach. When you chew, you're mechanically breaking down food into a mash that allows the enzymes in your saliva to make as much surface contact as possible to generate sugar. However, it's not even close to being as practical as koji for making sugar at a large scale for any form of mass consumption, let alone being largely accepted as a means for producing food. (However, there is a beer out there from Central and South America called *chicha de maize* that is made by chewing up corn and spitting it out to create an alcoholic beverage.)

Speaking of beer, brewing is one of the most widespread amylase enzyme applications that folks are enthusiastic about doing at home. Malted grain is the driver in beer making. The enzymes are generated by the process of sprouting grains to a specific point, stopped by dry heat. The amylase is suspended until the grains are steeped in hot water to make wort, the tasty sugar water that powers yeast fermentation and allows us to make beer. It's a pretty amazing process that in many ways is as natural to us as breathing. At the end of the day, you've been using enzymes to make your food more delicious since you were born. We feel that makes working with koji a natural progression.

Once you have sugar, you can power all the fermentation processes that you want to put to work for you. Let's walk through the life cycle of natural fermentation to cover how we get to each one. Overall, fermentation is driven by microbes both in our environment and in direct contact with the food, which are fed by added sugar or sugars inherent in that food, and sometimes controlled by salt. The basic cycle starts with *Lactobacillus* (and other beneficial bacteria generating lactic and other organic acids) and yeast feeding on sugars to produce alcohol. This can be followed by *Acetobacter* converting alcohol to acetic acid to give us vinegar. Acidic and alcohol environments are inhospitable to many pathogenic microbes, so this process naturally keeps our food safe when we follow specific guidelines, and the results are delicious.

The fermented foods we all know contain very specific ingredients and microbes. They were developed as a result of standardization for a known outcome driven by survival and now by popularity. People take for granted how much goes into maintaining the same flavor and texture for a fermented food no matter when you buy it. For example, commercial yogurt is made by inoculating milk with a very specific combination of *Lactobacillus* cultures, following a very tightly controlled process to yield a known product. It is

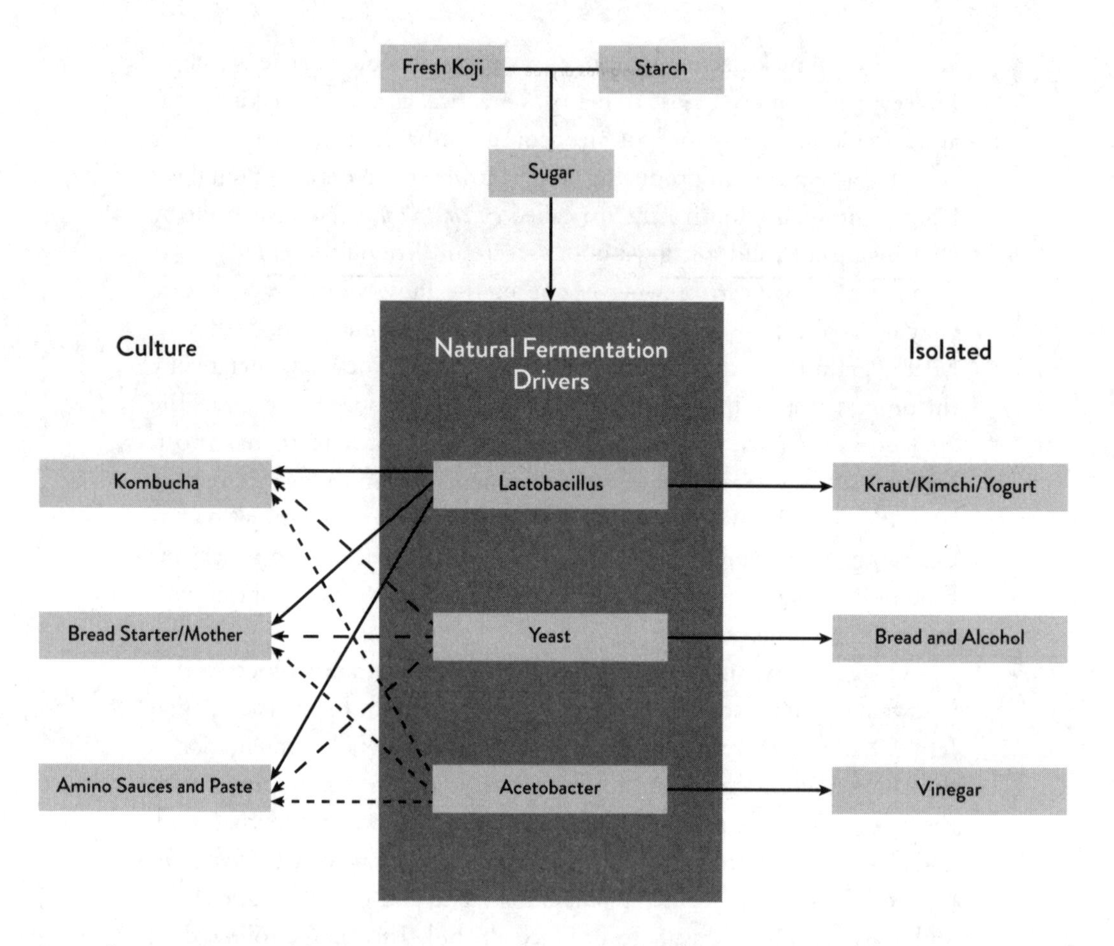


Sugar fermentation versatility map.

important to know that not that long ago, ferments were not as controlled. There were guidelines that kept people from getting sick, but ingredients varied based on what was on hand and there was much less control over the environment. Products would be used when they tasted right or—more important—when they were needed to nourish. Guidelines were much less specific and allowed more freedom for cooks to discover. While the amazing fermented foods we ended up with each have a very specific recipe for a reason, they are the results of a more free and natural process that was

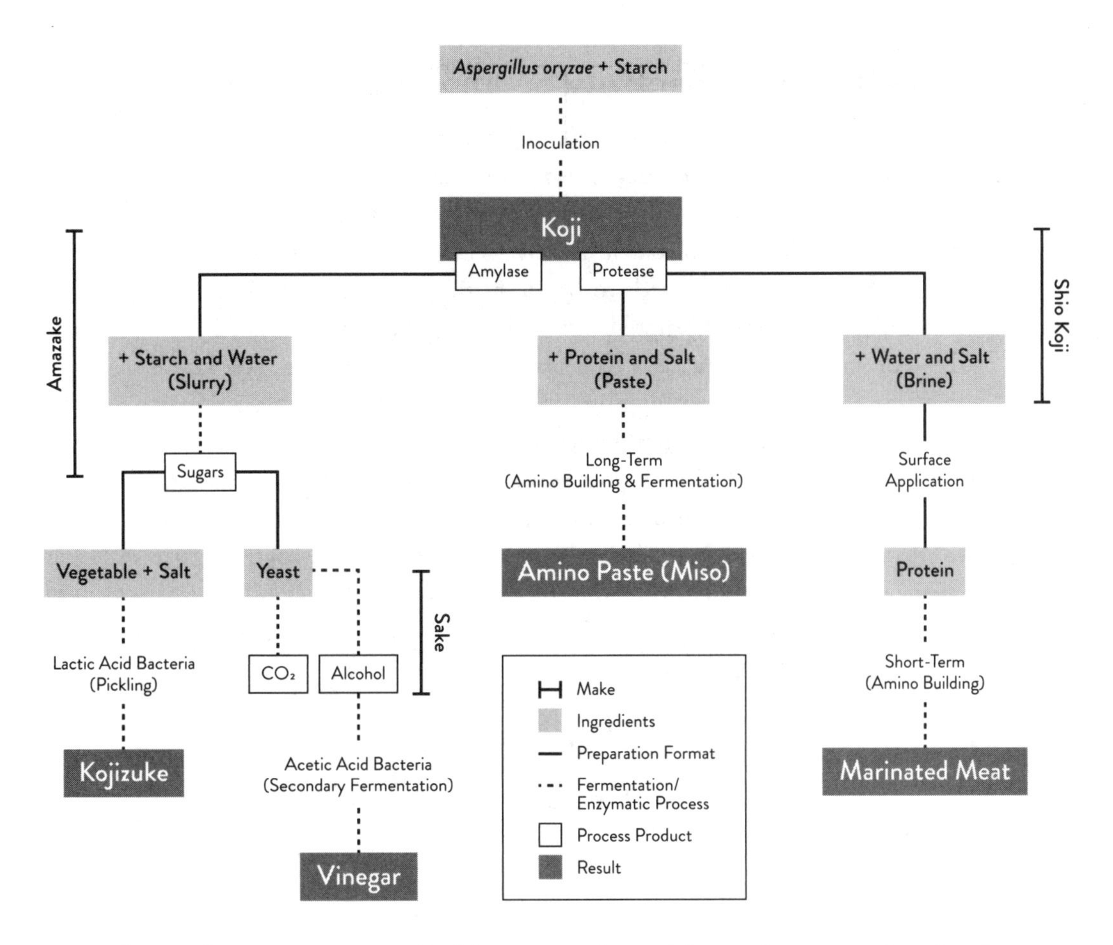


Master koji applications map tying all of the major makes and driving factors together. *Based on a flow chart illustration by Matthew Claudel.*

driven by a combination of survival and experimentation. Now consider all of the ingredients you have available to you to put practically any food through a natural fermentation process to see what happens.

Now that you understand how koji is so interconnected with making proteins more delicious, generating sugar, and fueling fermentation, the master koji applications map ties all of the major makes and driving factors together. We hope this helps solidify your understanding of how koji has the potential to make practically any food you eat more delicious.

## Why Is Interconnectivity So Important?

When you have been making food for many years, you start to see relationships among all the techniques and methods that you're practicing. You make connections that allow you to operate more efficiently, then you strip your processes down to the bare essentials to improve your execution. This understanding ultimately leads to freedom: the freedom to cook, create, manipulate, and combine ingredients in the ways that suit your needs. For those of you reading this who have classical culinary training, you know that this freedom is not something to take for granted. Culinarians can sometimes be militant in their desire to uphold what they view as classical tradition. In turn, they forget that those before them, while working to codify techniques, were also looking for ways to give cooks the freedom to create the best food that they could. Knowledge is a powerful thing that when used correctly can bring about the greatest of results.

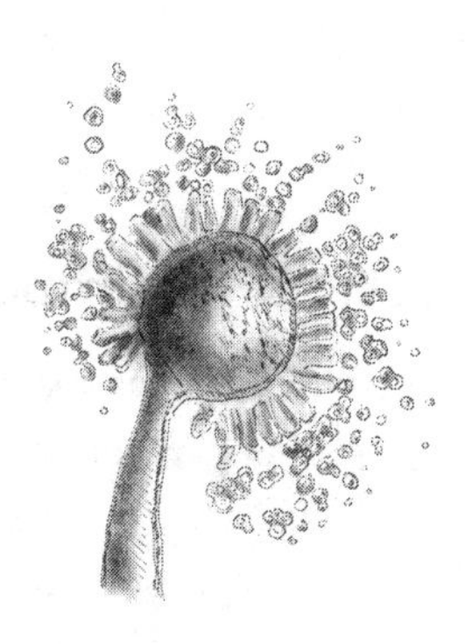

CHAPTER 4

# How to Grow Koji

Growing koji, or any filamentous mold for that matter, is very simple. For the most basic approach, all you need is a substrate, a container, spores, and a warm spot. But the most important and often most overlooked thing that you need is a *proper mind-set*. We know this may sound like some sort of transcendental new-age hipster statement, but let's face it: The most common interactions most of us have had with mold have been with either the putrid pile of leftovers lurking in the depths of our refrigerator or the panic-inducing black stains found inside our homes. These associations have led us to associate mold with spoilage and disease. While this is true for many molds, the ones used in food production are the exact opposite. These creatures are our allies in the kitchen; they not only make our foods delicious but are also responsible for making them safe. To a great extent we could make the valid argument that life as we know it wouldn't exist without these magical organisms.

There are many foods made with filamentous molds. *Penicillium nalgiovense* is what gives most styles of European dry-aged cured meats their characteristic white powdery coating, while *P. camemberti* and *P. roqueforti*

Rice plant. *Illustration by Max Hull.*

allow us to turn milk into Camembert and Roquefort cheeses. The Nordic yogurt-like food known as *viili* uses *Geotrichum candidum* in its production. In the Koreas they use *Rhizopus oryzae* to make alcohol, while Indonesians use *R. oligosporus* (and *R. oryzae*) to make the infinite varieties of tempeh that they are so fondly known for. The Chinese use *Monascus purpureus* to give *char siu* and other barbecued foods their stunning red colors. It should really come as no surprise then that filamentous molds should be able to find a home in your kitchen.

When we first started to entertain the idea of working with these molds in food production, we were hesitant. While our prior workings with fermented foods did make us more comfortable with the idea of working with mold in our kitchens, one cardinal rule about ferments kept eating away at us: "Mold is bad." When fermented foods such as sauerkraut develop mold growth on them, fearful chefs and food safety bureaucrats advise trashing the lot. Fermentation evangelist Sandor Katz repeatedly points out in his literature that this is unnecessary; all you need to do is scrape the mold off. But after adhering to the mold-is-bad standard for many years, it can be hard to train your mind to accept that not *all* mold is bad. This is the first hurdle you need to overcome when deciding to grow koji.

We often asked ourselves, *How will I know if this is the mold that I intended to grow?* And, *What if something went wrong? Will I sicken people?* You truly need to be comfortable with the fact that koji is a friend. There is always the possibility when cooking or producing a foodstuff that something can go wrong and cause a foodborne illness, but with a calm and collected mind-set you will be able to work safely within the risks and manage them without incident. This is something that others, with far less technological resources and scientific comprehension, have been doing for millennia. It can be done!

We are convinced that growing koji is a very intimate and personal experience just as all other aspects of gastronomy are. Feel free to put your personal touch into your koji as you become more comfortable with it; as much as our guidance is here for you to work with, don't feel that it is written in stone. This is exactly the train of thought that led us to develop some of the methods and techniques in this book. As we set out to work koji into our lives, some things deeply resonated with us, and others were simply lost in translation. Should you find that certain aspects of the growing process are either a benefit or a distraction, then leave them alone or optimize them to suit your needs. We fully encourage and support you in your quest to incorporate koji into the culinary adventures you find yourself embarking on. This especially holds true for growing koji. We are confident that you will find one of the following setups to be conducive to your wants and needs—but if not, go ahead and develop something that works for you. As with all scientific and creative endeavors, innovation and exploration are nothing without pushing boundaries and taking calculated risks.

As we discussed in chapter 1, koji is a living organism. It wants to live, survive, thrive, and reproduce, and it will do so as long as you provide it with a proper environment. *You don't grow the koji; it grows because it wants to.* Think of yourself as a farmer or a shepherd of sorts. Your job is to tend to your fields by keeping them hydrated, tidy, and free from unwelcome guests. This is exactly what farmers do for the fruits and vegetables in their fields and their animals out to pasture, and it's exactly what you need to do for your koji.

As we've previously stated, *Aspergillus oryzae* is a mold that proliferates in a tropical environment. Think of a hot and humid day to get a feel for what makes it thrive. Most everyone has experienced mold growth on bread that's been left out, especially when it is held in a plastic bag. Interestingly enough, one of the most common molds found on bread is a species of *Aspergillus*. Unfortunately, it is not as wonderful as the one we love, and yes, it will make you sick. It is the nasty stuff that makes folks toss out food all the time. The wonderful thing about our frequent encounters with mold is that it means that *it's easy to create favorable conditions*.

Spores, the mold equivalent of a plant's seeds, allow it to spread and proliferate everywhere. They're part of our environment and in the air all around us. The lightweight nature of these microscopic particles allows them to become airborne and be carried long distances on a light breeze. Spores are quite robust when it comes to survival. They can lie dormant for years until the right conditions come along.

The best way to start thinking about growing koji is to relate it to another beloved fungus microbe we harness for preservation: yeast. We all have a relationship with the amazing products that come from the wide range of fermentation processes that yeasts afford us: bread, beer, wine, vinegar, kombucha, et cetera. The key factors for creating alcohol are providing nutrition and providing the appropriate environment for the yeast to thrive. Most folks are aware of feeding a starter/mother for bread and pitching yeast into grape juice for wine. There is an optimal temperature range and a maximum to stay below so the microbe is not killed. Below that temperature range, the yeast is dormant and doesn't create the leavening or alcohol we're seeking. The nice thing about yeast is that the environment in our kitchens allows it to thrive. It's not that easy when it comes to koji, but it's not *that* much harder.

## Selecting Your Spores

One of the most common questions that we are asked is which spores to use when making koji. The traditional answer is straightforward, thanks to selective breeding of *koji-kin* (spore) producers. All you have to do is use the specific spore designated for the ultimate use, whether it be miso, soy sauce, sake, or pickles. Each type has been optimized for enzyme production on the specified medium, and it's insane how many different types are produced. Given how tuned in Japanese cooking and overall craftsmanship are, each koji spore has been precisely customized to its intended purpose. Basically, all koji types emphasize one of two enzymes: *protease* for protein breakdown to produce amino acids, and *amylase* for carbohydrate breakdown into sugars.

When both of us started this journey, we got hold of a bunch of different spores, made a variety of koji, and then used them to make the food products they're intended for. As we went through this stage of experimentation and education, we questioned the necessity of using each specific type. Because for most kitchens, managing a variety of spores is not practical.

We found that the light rice koji-kin that is customized specifically for short-term fermented foods (like sweet miso, amazake, and pickles) actually works well for every other application. Functionally, growing koji with this one type of spore gives you all the enzymes you need for breaking proteins down into amino acids for depth of flavor. We have never experienced a lack of umami in any application, from a shio koji marinade to an aged miso. We therefore found it easier to manage all our ideas and makes as we dove deep into the rabbit hole of possibilities if we stuck to just one spore variety for all our experiments.

There's one important consideration when it comes to universal use of light rice koji-kin, however. When you grow it on rice, the koji becomes sweet as intended, so you need to be aware of how that influences your short- and long-term applications. This will be explained in detail in chapter 5.

To be clear, we are not telling you to *only* stick to one spore type. Hundreds of years of breeding various kojis for specific results does have its advantages. If you are planning to optimize a very specific product, choosing the right koji-kin makes sense. There are also other flavor considerations when you're using koji types that produce acidity; you will learn more about this later in this chapter.

For most of us, the company GEM Cultures is where to start when it comes to purchasing koji-kin. They have a nice selection of all of the main applications a beginner would be looking for, and that range of products can sustain an experienced koji experimenter for quite a while. (Again, we used the light rice koji-kin for years before we started playing with others.)

GEM Cultures offer two size options: starter kits and commercial packs. Per their instructions, the starter kits only yield 5 pounds (2.3 kg) of koji, which is just enough to get a few trial batches under your belt. In contrast, the commercial pack yields 440 pounds (200 kg) of koji, which usually takes us a couple years to go through with extensive making and sharing.

## Spore Specifics

Commercial packs of spores are delivered in hermetically sealed pouches. Depending on the strain or species you buy, each pouch will be a different weight. Over the past several years, we've come up with a method of dispersing and using spores that works predictably every time.

Dispersal is the method used to put the spores into a workable ratio that allows you to use the perfect amount every time. You literally disperse, or mix, the spores into another medium such as all-purpose flour or rice flour. Dispersal also weighs down the spores and keeps them from floating away into the air you are breathing, which can be hazardous. As we have previously discussed, the spores of fungi are so small that a single one on your fingertip would be invisible to the naked eye. It would take tens of thousands of spores clustered on your fingertip for you to be able to see them. Their diminutive size means that they need to be spread out so that you don't use considerably more than you would ever need to for any given batch of koji.

When you receive your pouch of spores, you should first check to see that the room in which you intend to open them has no active air current. Turn off any ceiling and exhaust fans that may be present and close any doors or windows nearby. This is extremely important as a matter of safety. A disease called *aspergillosis* can be contracted by people that constantly are exposed to specific species of *Aspergillus* molds and their spores. Forms of this disease range from allergic bronchopulmonary to chronic pulmonary to cutaneous. Essentially what happens is that when certain species of *Aspergillus* enter your body, they think it is a food source. The spores can settle in your lungs, in nasal passages, or even on your skin and start to grow. Some of these species produce toxins called secondary metabolites that can be fatal. It is important to note this is of no concern with the species and variants that we use in food production! The spores themselves are so tiny and so many of them can compact in your respiratory system that they will actually calcify and cement in place. We don't want to scare you, but we feel it's important to address these issues. Any of the species or variants that you source from an approved and certified spore producer have been microscopically analyzed to ensure that they are food grade and not pathogenic species or potentially contaminated with other worrisome little buggars. The people who work in the grow rooms at approved production facilities wear respirators and other protective clothing to protect them from the copious concentrations of spores that they are in contact with. Short of intentionally inhaling a pouch of spores, you will never be subjected to the concentrations these koji producers are exposed to and only need to take minimal precautions when dealing with spores that haven't been dispersed. In short, if you wear a bandanna or respiratory mask when working with concentrated spores, you will be just fine.

If you're wondering whether you can just grow your own spores like you would keep a vinegar mother or sourdough starter, the answer is yes, but you shouldn't. As we noted above, the companies that produce spores submit each batch to be analyzed to ensure that they have a high level of purity and that no deadly pathogens have contaminated them. Several layers of regulations govern those who produce spores for use by others. Anyone who has grown fruiting fungi understands that spores that aren't produced in sterile environments under extremely controlled conditions will most likely become contaminated by a host of other microscopic organisms, spelling doom for the intended growth in a domesticated, cultivated setting. This holds true for all fungi, from mushrooms such as maitake to filamentous molds such as koji. Yes, the risk of being contaminated with a potentially toxic pathogenic species of *Aspergillus* or other mold is extremely low, but it can happen. Why gamble when there's a better, safer way? The koji industry in Japan has had some serious setbacks over the past ten years. There are fewer and fewer people growing their own koji and therefore fewer people buying spores. By buying spores from a verified source, you help keep the koji industry, one that has been largely family-owned and around for centuries, relevant and operating. You also help keep the available species and species variants of koji biodiverse. When people purchase koji spores from producers, the companies keep investing in breeding variants of species and species that are optimized to perform certain tasks. These tasks range from the high protease activity best suited for amino sauce and amino paste production to high lipase action that produces kojis with incredible aromatic qualities that outshine even the most fragrant of flowers. Without this investment in innovation, the koji industry would surely begin to mirror what happened to the apple industry, a disaster of Red Delicious disgustingness, and the types of koji you would have access to would surely shrink.

Take it from us and purchase your koji spores. You eliminate potentially deadly contamination risks, keep an awesome ancient industry thriving, and protect important biodiversity and species evolution. On top of that spores are relatively inexpensive, especially given the yield you obtain in inoculated substrates. A roughly 40 grams (1.4-ounce) packet of spores will run you about $35 USD, shipping included, and produce hundreds of pounds of koji.

## SPORE DISPERSAL INSTRUCTIONS

The dispersal ratio that we work with and recommend is 1:10, which means that for every 1 gram of spores you need 10 grams of medium with which to disperse it. Our choices for dispersal media are rice flour and all-purpose wheat flour. If you want to keep the foods you make from your koji 100 percent gluten-free, then use rice flour.

Once you've dispersed your spores, the ratio to then culture a substrate—rice, beans, or anything else you are attempting—is 1 gram of dispersed spores per 1,000 grams of cooked substrate.

The procedure:

1. Grab a mason jar or other container with a tight-fitting lid (a 1-quart/liter mason jar works well).
2. Using scissors, carefully open the packet of spores. While pinching shut the opening you just made, lower the packet into the jar all the way to the bottom before allowing the spores to fall into the jar. Wait until the spores visibly settle before you remove your hand and the packet that the spores were in.
3. After you remove your hand with the empty packet, you can begin to add the flour. Gently and carefully add just a little at first, enough to cover the spores and weigh them down. You want to keep them from becoming airborne, so be gentle.
4. Once a layer of flour is covering the spores, you can add the rest of the flour.
5. Tightly cover the jar and shake it for a few minutes so that the spores evenly disperse within the flour. The spores will turn the flour a color similar to their own. For example, *Aspergillus oryzae* will turn the flour a light shade of army green, while *A. luchuensis* var. *awamori* will turn the flour a light charcoal color.
6. Once your spores are evenly dispersed, they are ready for use.
7. If you're not using them right away, store the spores in a cool dark place with the lid on tight. We prefer to store them in the freezer. We have stored dispersed spores for as long as eighteen months, and they were just as viable as ones fresh out of the package. The goal is to keep your spores away from conditions that will allow them to grow until you intend to use them. It's never a bad idea to also source desiccant packets to keep in your sealed jars. They'll act as an extra bit of insurance by keeping moisture from condensation away.

## Tradition Is Paramount

The traditional steps for growing rice koji, as outlined by the Japanese, do differ slightly depending on your desired results. Sake makers will mill their rice to remove the outer layers that contain fats and proteins (which are mostly undesirable in an alcoholic drink), whereas miso makers will abstain from milling because they want more protein in their miso. These differences also translate to how you grow koji. For example, sake makers must take great care to ensure that their koji is nothing short of perfection, for sake is a drink that is to be enjoyed on its own, enjoyed for every minute nuance. When drinking sake you can directly feel the koji from which it's made—the texture and body that the proteins in the rice endow, the highly enticing aromatics lent by fats, and the balanced sweetness extracted from its starches. This is why the traditional guidelines for growing koji are so strict and stringent. If even the slightest deviation is made, then a product is not worthy of attaching the brewer's name. No brewer wants this, so they follow the steps in a way that can be described as "rigidly artistic."

The traditional setups for miso and shoyu, on the other hand, while they emulate the methodology of the sake maker, allow a considerable amount of leeway. Miso and shoyu makers create foods that are intended to be a part of a symphony, as opposed to sake's solo act. Should miso makers overgrow their koji to the point of it sporing, for example, they need not fret too much, as the dank and musty flavors and aromas that spored-out koji carries are masked or lost during the time it takes for the miso to mature. This also extends to culturing koji on meat, vegetables, and dairy products, where some of those flavor profiles are desirable. Facilities that produce sake are often sterile, with workers wearing stark white garments. Compare that with the facilities where barrel-aged miso and soy sauce are made—rustic environments that lend the additional microbes for long-term fermentation.

## Making Basic Rice Koji the Traditional Way

Let's walk through the basic rice koji making steps—both the traditional methods and what's worked for us.

1. **Source the proper rice.** Koji prefers the starch amylose, which is found in higher concentrations in long-grain rice. There are three main grades of rice grains: short, medium, and long. Generally speaking long-grain varieties contain higher ratios of amylose to amylopectin, whereas short-grain varieties are the opposite and contain a higher ratio of amylopectin to amylose. While brown rice does contain a higher density of micro- and macronutrients for koji, it is mainly reserved for savory applications such as amino pastes. Long-grain white rice is the best all-purpose substrate. We love the koji we obtain from jasmine rice. This isn't to say that you should only use jasmine rice; it's just what we generally prefer. You may also have a preference for organic versus non-organic rice. Simply put, choose your rice based on what's important to you and the results you want to achieve.
2. **Wash the rice.** This is done to rinse the residual starches left on the rice after it has been milled. During the milling process used to obtain white rice, the hull, bran layer, and embryo are all removed. As this is done a considerable amount of pulverized starch is left on the grains. The presence of this excess starch is very evident if you place your hand into a pile of raw grains. When you remove your hand, it will feel dusty and chalky. The residual starch can cause your rice to stick and

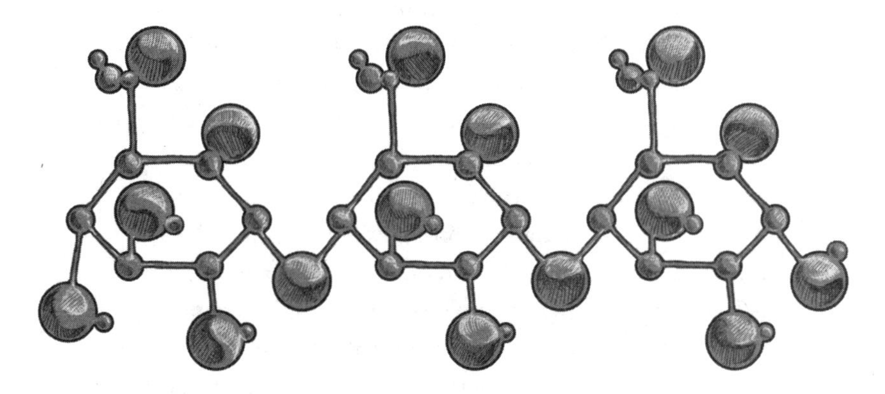

Amylose molecule. *Illustration by Max Hull.*

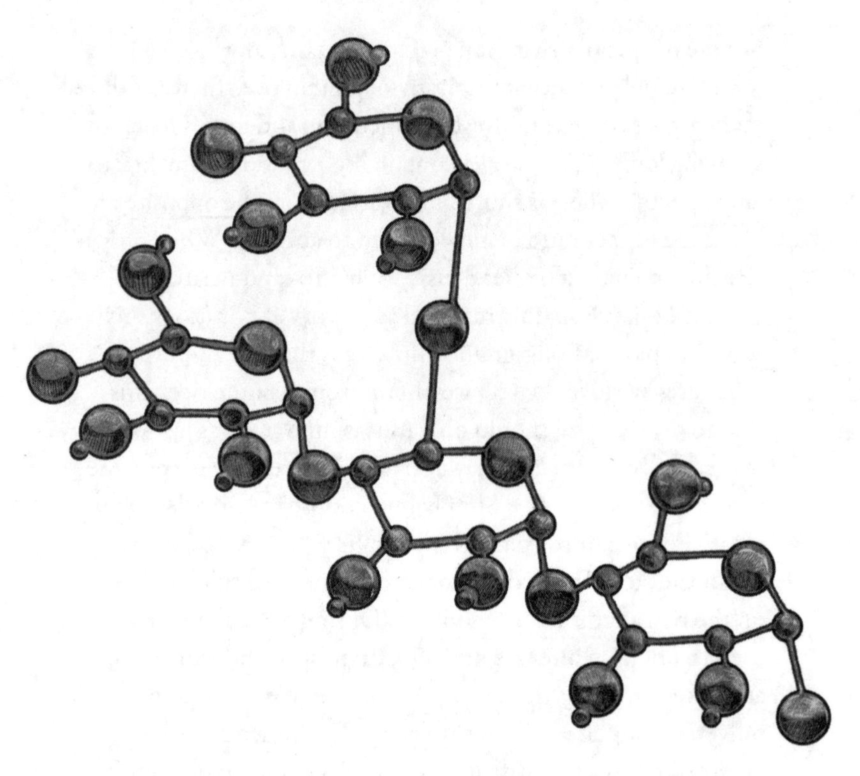

Amylopectin molecule. *Illustration by Max Hull.*

clump when it is cooked, which is not conducive to growing a strong sake koji, for example, as it needs proper air exposure. For this reason, you must wash the rice under cold running water until the water is crystal clear and no longer clouded by free starches. The washing process can take 30 minutes or more, depending on the variety of rice and how clean you want it to be. This is most important factor driving the base flavor of a sake. For most other applications, however, washing doesn't need to be as diligent—it's even okay to skip it altogether.

3. **Soak the rice.** The rice must be soaked at ambient temperature for at least 6 hours. This is done to partially hydrate the rice so the appropriate gelatinization of the grains can be achieved during steaming. This is an important step that you must not skip.

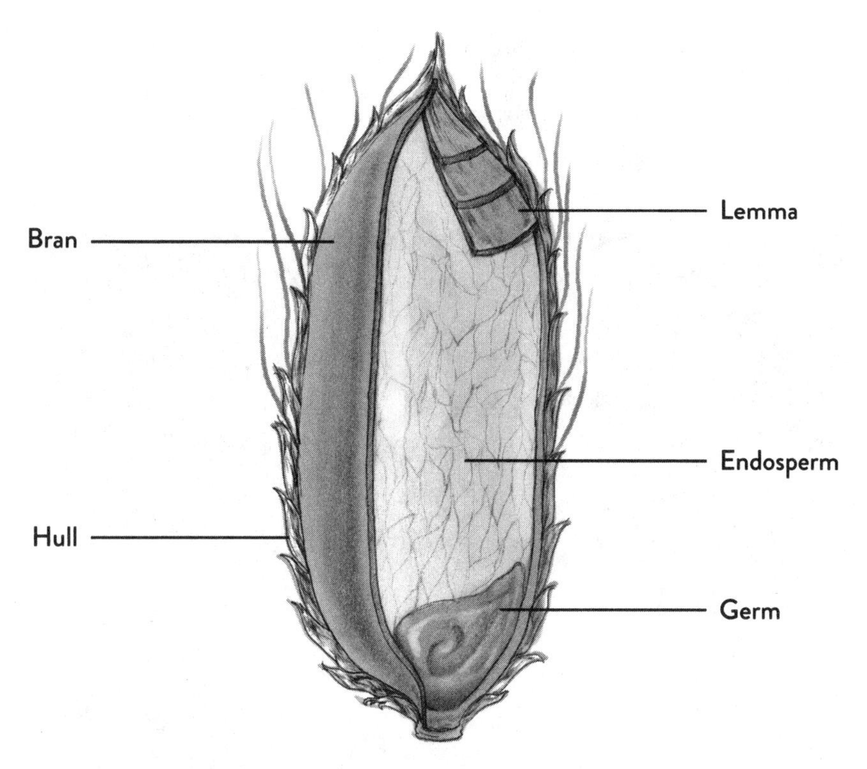


Rice grain anatomy. *Illustration by Max Hull.*

4. **Steam the rice.** Traditionally steaming was done in a bamboo steamer set over a pot of boiling water. First, line the steamer basket with a lint-free cloth to prevent the rice from falling through the wooden slats. Then drain the rice from its soaking liquid and put it in the basket. Fill other steamer baskets of the same size accordingly and stack them on top. Place the lid on the basket and let the rice steam for 45 to 75 minutes. The goal is to cook the rice to the point you'd consider pasta to be done (al dente). The Japanese concept of *gaikou-nainan* encompasses this—that moment when the surface of the cooked rice is hard and the inside is soft. This forces the mold to focus on growing *into* the rice instead of just *on* the surface.

5. **Cool the rice.** Spread out the rice evenly onto a large surface and sometimes fan it to allow it to cool. For those of us making smaller batches, hotel/baking pans work well. In general, any food-safe pan or tray that will fit into your incubator will get

Rice ready in a bamboo steamer. *Illustration by Max Hull, based on a Patrick Soucy photograph.*

the job done. An evenly filled tray should have no more than a 1½-inch-high (3.8 cm) bed of grain. Koji will start to die off if it is exposed to temperatures above 115°F (46°C) and will be completely dead at around 130°F (54°C). It prefers a temperature range between 70°F (21°C) and 95°F (35°C). The cooling process helps to dispel excess moisture and allows the gelatinized starches in the rice to set in place. Koji makers can gauge whether the rice is cool enough just by touching it. Or you can use an accurate probe or IR gun thermometer.

6. **Inoculate the rice.** Koji spores were traditionally applied using small handheld bamboo shakers similar to salt shakers. The koji maker would sprinkle the spores over the rice in such a way that an evenly dispersed cloud would rain down onto the rice. On a small scale we've found that putting the amount you need on a small spoon and lightly tapping as you move over the rice bed works just fine. The ratio we use is 1 gram (0.04 ounce) of spores to 1 kilogram (2 pounds) of rice.
7. **Mix the rice and spores.** Gently mix the rice and spores by hand to ensure that the spores are evenly dispersed and cover as much rice as possible. You want to do this such that the rice grains are neither crushed and broken, nor smashed together to form large sticky clumps. Koji needs exposure to atmospheric oxygen; if the grains stick together, no koji will be able to grow. A thorough mix for a few minutes is enough to disperse the spores.
8. **Mound the koji and cultivate the rice.** Traditionally, the inoculated rice is heaped into a large pile on a tarp, covered, and rested for a minimum of 6 hours. In your case, leaving the rice in a tray spread evenly for 12 hours works well. How the rice is specifically covered depends on the incubator you're using. (Those details are described in step 10, "Tray the koji.")
9. **Mix the rice.** After the first rest, mix the rice. This does three things over the course of the growing process: It aerates the rice, cools it, and spurs vigorous mycelial growth. As koji grows, it generates heat. The heat can swing so intensively that it rises above 130°F (54°C), thus killing the koji before it has time

to fully produce its mycelial mat. Without this step the koji would fail to thrive and then spoil and decompose. As the koji is mixed, the mycelium is broken into smaller pieces that in turn grow with more voracity. This, in some respects, is similar to the reasoning behind pruning a fruit tree.

10. **Tray the koji.** Place the inoculated mixed rice into square or rectangular trays made from Japanese cedar wood that have been lined with lint-free cloths. Instead of spreading it into a flat layer, form the rice into numerous ridges or mounds that rise and fall over the surface. These ridges can be linear and parallel, or you can arrange them in a circular pattern. This step is an art, meditative in the same way as a thoughtful Zen garden. It's also practical, however, in that these ridges and mounds increase the surface area of the rice and expose more of the grains to air. The trays are covered by lint-free cloths to help maintain high humidity, then stacked and allowed to rest. The cloths are called *tenugui* in Japanese and are made of a very finely woven cotton fabric. They do not easily fray or shed lint that can become stuck to the koji. You can easily find lint-free cotton cloths at many retailers.

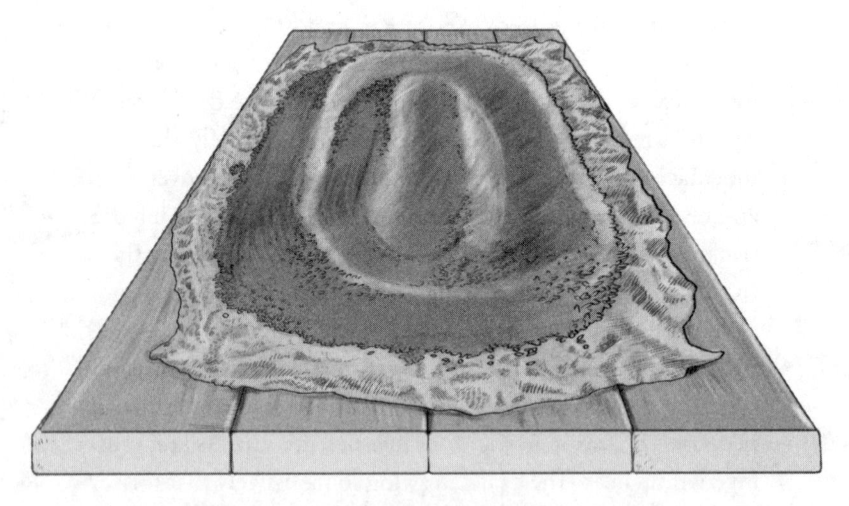

Mounding rice helps to cool it. *Illustration by Max Hull.*

11. **Mix and rest, mix and rest.** The mixing and resting cycle is executed a number of times until the koji has fully bloomed and grown deeply into each grain of rice. You can break grains of rice to see how deep the hyphae of the mycelium have penetrated. (The closer to the center of the grain, the better.) We've found that mixing and mounding every 12 hours until you hit 48 is sufficient.
12. **Give the koji a final inspection.** Carry out a final inspection of the koji rice. Aromatic and visual cues will let you know exactly when to harvest the koji and use it as needed. The koji will become extremely fragrant with an aroma that many people describe as a mix of fresh fungus, chestnuts, honeysuckle, champagne, and tropical fruit. The main visual cue is a stark white fluffy mycelial mat that has completely and tightly bound the grains of rice together. Once these two cues are apparent to your standard, the koji is ready. From start to finish this entire process averages between 2 and 4 days.

---

A similar methodology can be used to grow koji on any bean or other grain or seed that is high in starch. When we first started growing koji, we solely used *Aspergillus oryzae* to inoculate our substrates. As time went on and we became more comfortable, we started to use a wide variety of not only different strains of *A. oryzae* but also different species of koji such as *A. sojae* for legumes and *A. luchuensis* var. *awamori* for dairy. No matter the strain or species of koji, however, the steps to securing, preparing, inoculating, and growing koji on a given substrate are the same.

## Modern Approaches to Growing Koji

Modern approaches to growing koji employ a wide variety of equipment and setups ranging from a version of the traditional Japanese koji growing cabinet to nothing more than a baking sheet and plastic wrap. In this section we'll discuss the options for various price points and levels of commitment.

We will cover options ranging from absolute bare-bones needs to leveraging technology found in fine dining restaurants and some home kitchens. At the most basic, growing koji can be simply achieved with cooked grains kept under humid conditions, so very little is required. What's essential is setting up or adapting a controlled environment that can be sustained for a couple of days.

We've worked extensively with what we refer to as standard cooked rice, also known as boiled rice and pilaf rice. While this rice typically allows for less optimized koji growth, given the proper attention—and in some cases piggybacking koji with lacto-fermentation—will minimize this. We have always had plenty of enzymes; any slight souring isn't a problem for most applications—especially amino pastes and amino sauces. For most people—especially those trying to make enough koji to supply a busy restaurant—it's a beautiful thing to have room to be less fussy with the rice preparation. We use standard cooked rice at Larder every day as our primary substrate for growing koji to be used in most of our products.

The technique we developed for producing koji with standard cooked rice allows for ease of preparation, incubation, and final use. It all starts with selecting the proper rice—so, long-grain. (As we previously discussed, long-grain varieties of rice have more of the starch amylose, which is koji's preferred food.) We love using jasmine or basmati rice due to their fantastic inherent aromas and flavors. Economics are another strong incentive to use jasmine, as you can easily purchase a 50-pound (23 kg) bag for under $40 USD. That's less than a dollar a pound, which is very advantageous in a commercial food setting. Making and using koji requires a decent labor investment, so keeping the ingredient cost down is a huge plus. Once you have your rice, it's time to prepare it for inoculation, inoculate it, allow the koji to grow, and then use or store it. In the following steps you'll notice a few differences compared with the traditional process in the previous section—ways we have adapted the process to how our kitchens function and the ways in which we use koji. We'll discuss some of these differences and why we made the decision to change.

A major reason for some of these changes was that the processes were developed in chef-driven professional kitchens. Those of you who work in one will be able to, we hope, relate. Chef-driven professional kitchens are staffed with professional culinarians of varying degrees—everyone from the dishwasher to the executive chef—a majority of whom are new to cooking and gastronomy and are learning on a daily basis. This being the case, tasks often are delegated to individuals based on their skill sets. In many kitchens dishwashers and prep cooks are tasked with various duties such as peeling onions, potatoes, and garlic, picking and cleaning herbs, and preparing lemon wedges. Rarely are they asked to butcher a fluke, make a sauce, or create a fermented food. These are tasks that often fall to specialized workers within the kitchen, be they a saucier, a sous chef, or a butcher. While developing the techniques for growing and using koji in a kitchen such as this, our intent was for it to be simple. We wanted anyone handy to be able to take on the task should we be busy with other tasks. Due to this we eliminated what we could from the traditional growing process and kept it as approachable as possible. The great result of a simplified culinary methodology in a chef-driven professional kitchen is that once it becomes approachable for individuals who don't have the skills yet to execute a complex process on their own, anyone can follow it. And you don't have to be working in such a kitchen to embrace this simplified method. Anyone, professional and amateur alike, can embrace this process and work comfortably with it.

The following steps are essentially the same not only for all types of rice but also for all legumes, seeds, and anything else that you can get your hands on. All you really need to do is cook, inoculate, incubate, and then use. (Check the "Popcorn Koji" recipe on page 107 for some of the specifics you'll need when working with substrates other than rice.)

One thing to keep in mind with this method is that some of the rice on the bottom of the dish will be overcooked. This has both pros and cons. The major con is that you might be left with some mushy overcooked rice. The pro is that that koji can and will grow on the overcooked rice, and it's also perfect to use for making amazake (which we'll detail later on in "Amazake" on page 112).

1. **Preheat your oven to 350°F (176°C).** This methodology is based on European, specifically French, styles of cooking rice. The cooking is done in the oven rather than on the stovetop

or in a steamer. Place your rice into an oven-safe dish filled no higher than half the depth of the pan. We use a 6-inch-deep (15.2 cm), full-sized hotel pan, but you can use any vessel that will allow you to cook the amount of rice that you intend to culture. Even a stockpot will work.

Lay your hand flat on top of the rice and add enough cold water to just cover your knuckles. In this process we don't wash the rice. That isn't to say that you can't, but as we've stated this process is simplified for ease.

2. **Cover your dish and bake.** Once the rice is covered with water, cover the dish with a tight-fitting lid or foil and place it in the oven for 90 minutes, then remove it. Allow it to rest, covered, for at least an hour.
3. **Remove and break apart clumps.** Once the rice has rested, remove it from the dish and separate out the wet overcooked parts, saving them for making wet koji or amazake. The properly cooked rice should be gently broken apart so that it flows freely with no clumps present. This is very important and serves to ensure that your koji will grow beautifully. Clumping prevents airflow from reaching the grains. Breaking the rice apart allows for greater airflow within the rice as the koji grows. Grains that are separated also allow for better dispersal of the koji spores and make it easier for the koji to grow.
4. **Inoculate the rice.** Now that the rice is broken apart, it's time to inoculate it. First take the temperature of the rice; it should be no hotter than 100°F (37°C). Temperatures hotter than this can kill the spores. You can use a thermometer or simply rest your hand on the rice. If it feels hotter than your hand, wait a little for it to cool off to the proper temperature. Once it's cooled, you can sprinkle on your spores. Do this so they evenly rain down onto the rice. Gently mix and massage the rice to fully disperse the spores. Be gentle and use your fingertips. You don't want to crush the grains or smash them back into clumps. Carefully place the inoculated rice into a metal tray. Don't compact the rice, and don't let it get more than 2 inches (5 cm)

deep. This is important to allow the koji to grow undisturbed. Layers of rice deeper than 2 inches will need to be mixed as outlined in the traditional method. Layers less than 2 inches deep won't need to be mixed. They aren't thick enough to overheat. This is one of the major advantages to this method.

5. **Cover the tray tightly with plastic wrap.** Poke a couple of finger-sized holes in the plastic wrap and then incubate at the desired temperature, 80°F (26°C) for higher protease production or 90°F (32°C) for higher amylase production. The koji should be fully grown approximately 36 hours later, after which you can use it or store it for later.

## Storing Your Koji

There are multiple options for storing your freshly grown koji. Wrapped tightly in plastic wrap or in a zippered bag, fresh koji will hold just fine under refrigeration for around two weeks. In a vacuum-sealed bag you can store fresh koji in your refrigerator for six weeks. If you absolutely have to freeze it, vacuum-pack it first to remove as much air as possible. You can freeze koji this way for up to six months, but it will not be nearly as good as fresh. Our preference is to always use it immediately, especially given the time and effort you've put in.

Should you not have room in your refrigerator or freezer for your freshly grown koji, you can also dehydrate it. Gently dry the koji at 100°F (37°C) until it is brittle. If you dry it above 165°F (73°C), you'll denature the enzymes—they are proteins after all—and they'll more or less be useless. You could still use it for the sugars it has produced in fermented foods that rely on simple sugars, but once again the whole point to using koji in the first place is the active enzymes. Once your koji is dried, you can powder it and use it for a whole host of creative applications—as a seasoning or a thickener, for example—or you can leave it alone and use it for more traditional applications. Whatever you do, store it in an airtight container or package and keep it away from heat, moisture, and light. Stored this way it should be used within a couple of months.

In short, store your fresh koji in whatever way you are comfortable with and use it sooner rather than later. Always keep in mind that the enzymatic action of your koji decreases the longer it sits around.

## Trays

As we've discussed, *Aspergillus oryzae* needs oxygen to grow. The easiest way to provide that is to offer as much surface area exposure to the air as possible. Traditionally steamed grain is mixed in a large cloth-lined bed until it's cooled; then spores are added, and the grain is transferred to another cloth-lined bed to retain heat for preliminary growth, then finally moved to cedar trays to allow for heat dissipation in the final stages. The benefit of cedar is its insulative ability, which can keep the grain at a constant temperature. It also wicks water so it doesn't bead or pool. The mold that starts to reside in the wood after multiple uses is an added bonus. However,

A koji *buta*, the traditional Japanese koji growing tray. *Illustration by Max Hull of South River Miso's trays.*

unless you are a traditional koji maker producing at high volumes, cedar trays are tough to manage.

Today koji is made in kitchens all over the world using standard food service trays, typically hotel pans (stainless steel rectangular pans with walls 2 inches / 5 cm high) or sheet trays. They're readily available, inexpensive, and easily cleaned. Growing koji from start to finish in trays works well enough for the finest of fine dining restaurants. Honestly, any low-profile non-reactive vessel that will fit into your incubation chamber will work.

Remember that since you're working with food, cleanliness is key. The incubation conditions of koji are optimal for harmful food microbes. It's important that the incubation chamber and everything that goes inside it are cleaned to the standard you would follow for any food service vessel.

## Bare-Bones Incubation Needs

There are a few simple requirements to grow koji: (1) a mild heat source to hold the starch medium at a warm temperature for two days; (2) a level of ambient airflow to supply oxygen; and (3) a fair amount of humidity that does not allow water to bead on the grain bed.

The optimal warm temperatures required for incubation are 86°F (30°C) (protease to break proteins down into amino acids) and 95°F (35°C) (amylase to break starches into sugars). If you think a little about where you live and kitchens you work in, we're sure you can think of a few spots that are warm and cozy: on top of a refrigerator, inside your oven with just the light on, in the basement next to the furnace, et cetera.

The best way to maintain the environment is to start in a place that is already as close as possible to ideal and controlled. Most places that we go these days have their thermostat set to a comfortable temperature for humans, so getting it up to the appropriate temperature for koji doesn't take much energy at all.

## Water Bath Method

One of the simplest and least expensive incubators you can create with basic equipment you may already have at home is a contained and lightly heated water bath. The two keys to this system are: (1) leveraging the surface area of exposed water to create the humidity required through evaporation; and (2) water offering the heat distribution and capacity to maintain the temperature. Insulation of the incubator is another important factor to consider depending upon the environment you are working in. If the room is controlled fairly well with a thermostat set to 65°F (18°C) more or less, you probably can get away without it.

Here's what you'll need to build the system. These needs are described as broadly as possible so you can meet them however you see fit:

- A large watertight container and a way to cover it.
- Non-reactive and food safe tray(s) that will fit inside the container.
- Clean, lint-free, food-safe cloth(s) to cover the cooked grain.
- Lightly heated potable water that will maintain the desired incubation temperature.
- A way to suspend the trays so they are not in contact with the water bath.
- A method to assist in humidity generation is helpful, but not always necessary.

## SMALL-SCALE WATER BATH

Here's how to put together the small-scale starter incubator we recommend when folks are first getting into making koji. It will yield 1 pound (0.5 kg) of koji. The key to this system is the immersion circulator that maintains the desired temperature and agitates the water to assist with generating humidity.

### EQUIPMENT

One 8-inch-deep (20 cm) full-sized Lexan/polycarbonate hotel pan
One home immersion circulator
Four 8-ounce (236 ml) mason jars or plastic pint deli containers
One 9 × 13-inch (23 × 33 cm) rectangular cake pan with prepared grains
One kitchen towel
Plastic wrap

Fill the Lexan pan with warm water to the recommended level for your immersion circulator and below the height of the jars you are using. Install the immersion circulator on one short side of the water bath. Fill all four jars with water to the same level as what is in the Lexan to keep them from floating. Place them in a rectangular pattern in the water so that they will both support the cake pan and keep it away from the immersion circulator. Set the prepared grain tray on the jars. Place a clean, damp towel on top of the grain bed. Cover the top of the Lexan with plastic wrap. Set the immersion circulator to the appropriate temperature.

## LARGE-SCALE WATER BATH

This incubator is the heart of all the koji Rich has been making since his first batch years ago. He learned all about this setup and the basics of koji making from Branden Byers, author of *The Everyday Fermentation Handbook*. The beauty of the system is that it uses a large cooler as the chamber, which offers everything that this incubator needs: a watertight vessel for the bath, a cover to hold humidity, and insulation to maintain temperature. The only additions required are a heat source and a humidity generator, both of which can be created with inexpensive aquarium components you can pick up at any major pet store.

Note: We have recently started adding salt to the water bath as a precaution for keeping harmful microbes at bay. All you have to do is make a minimum 3 percent brine, adding salt by weight to the water. Those of you who have home charcuterie setups are already aware of this way to add a touch of humidity to your home rigs.

### EQUIPMENT

One large cooler that will fit a hotel pan inside
One 3-foot (1 m) minimum length of aquarium airline tubing
One air stone, 4-inch (10 cm) disk or 8-to-12-inch (20–30 cm) bar
One 10-gallon (38 L) aquarium air pump
One 150-watt adjustable submersible aquarium heater
Four 8-ounce (236 ml) or 16-ounce (473 ml) mason jars

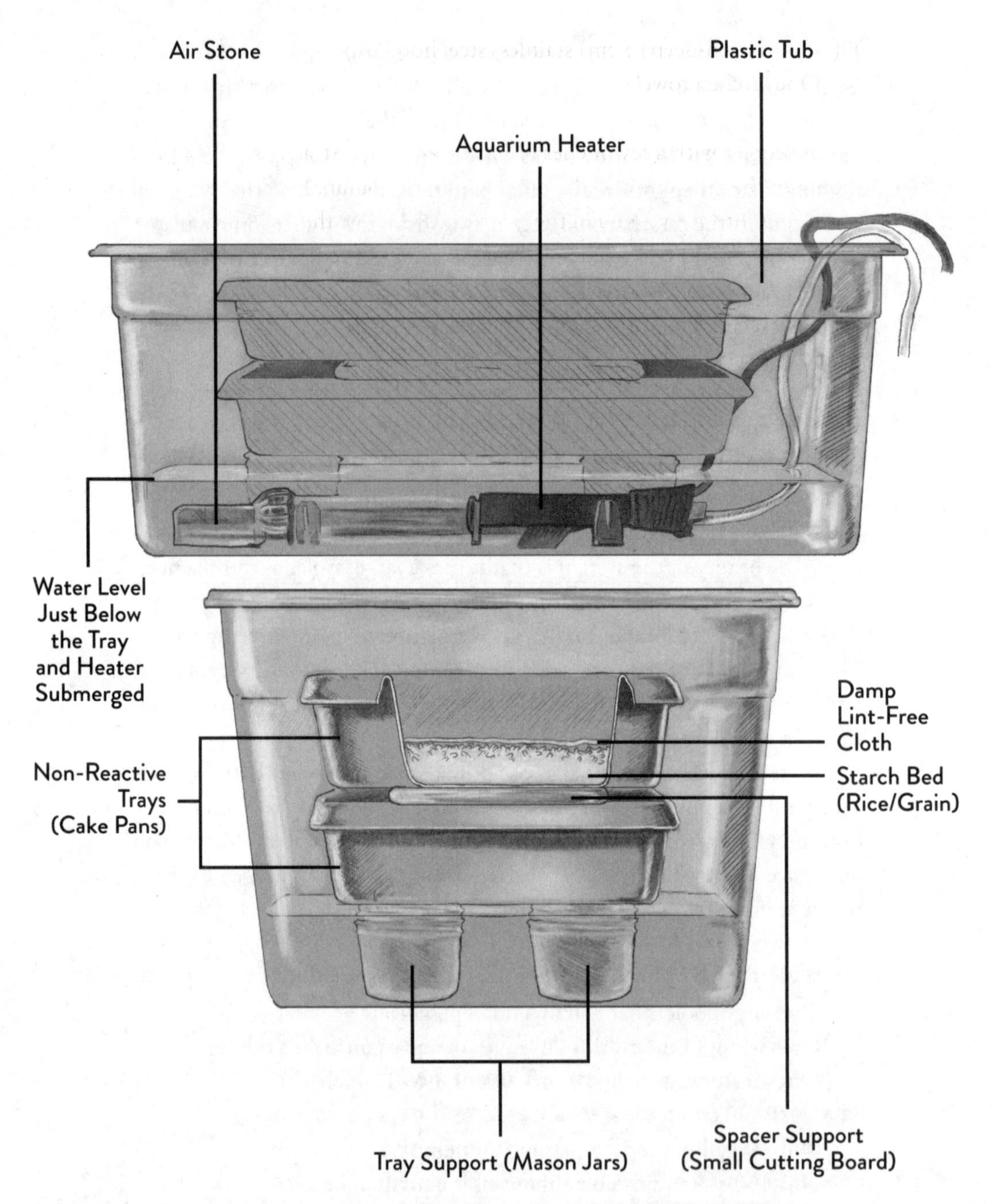


Water bath incubator with aquarium heater and two trays. *Illustration by Max Hull.*

One 2-inch-deep (5 cm) stainless steel hotel pan
One kitchen towel

Fill the cooler with a few inches of 3 percent brine. Attach one end of the tubing to the air stone and the other end to the pump. Place the air stone and heater into the water. Fill the jars with water past the level of the water bath to keep them from floating. Place the jars in a rectangular pattern to support the hotel pan. Set the heater to 86°F (30°C). Plug in the pump and heater. Cover the hotel pan containing the prepared grain with a damp kitchen towel and let it sit on the jars above the water. After the pan is in place, run the tubing and heater cable to the front corner. This will allow the cover to close enough to maintain the heat and humidity so you don't have to modify the cooler. If you would like to close it completely, it's easy to cut a slot at the top of the cooler with a box cutter.

## Dehydrator or Dry-Heat Incubator Method

A dehydrator is a controlled dry-heat incubator that can be set to the appropriate temperature for koji. While it presents a challenge in that it eliminates the humidity that mold needs to grow, we discovered long ago that all you need to do to overcome this is cover the tray/container. You can thus create humid conditions with the inherent moisture in the grain bed itself without adding any water. The only other consideration is allowing for airflow, which just requires poking some holes through the cover.

When looking for a dehydrator, be sure to find one that can hold temperatures in a lower range. Many of the available makes and models have a temperature range that starts above 100°F (37°C) and ends at 165°F (73°C), and as we already stated these temperatures are too high. The recommended work-around with these incubators is to leave the door open to varying degrees or even take it off, and to use a thermometer to help you figure out how to keep the temperature in the preferred range.

At Larder we use a commercial dehydrator that can hold temperatures as low as 50°F (10°C) with fairly consistent accuracy. Check out the resources section at the back of the book for more info about it.

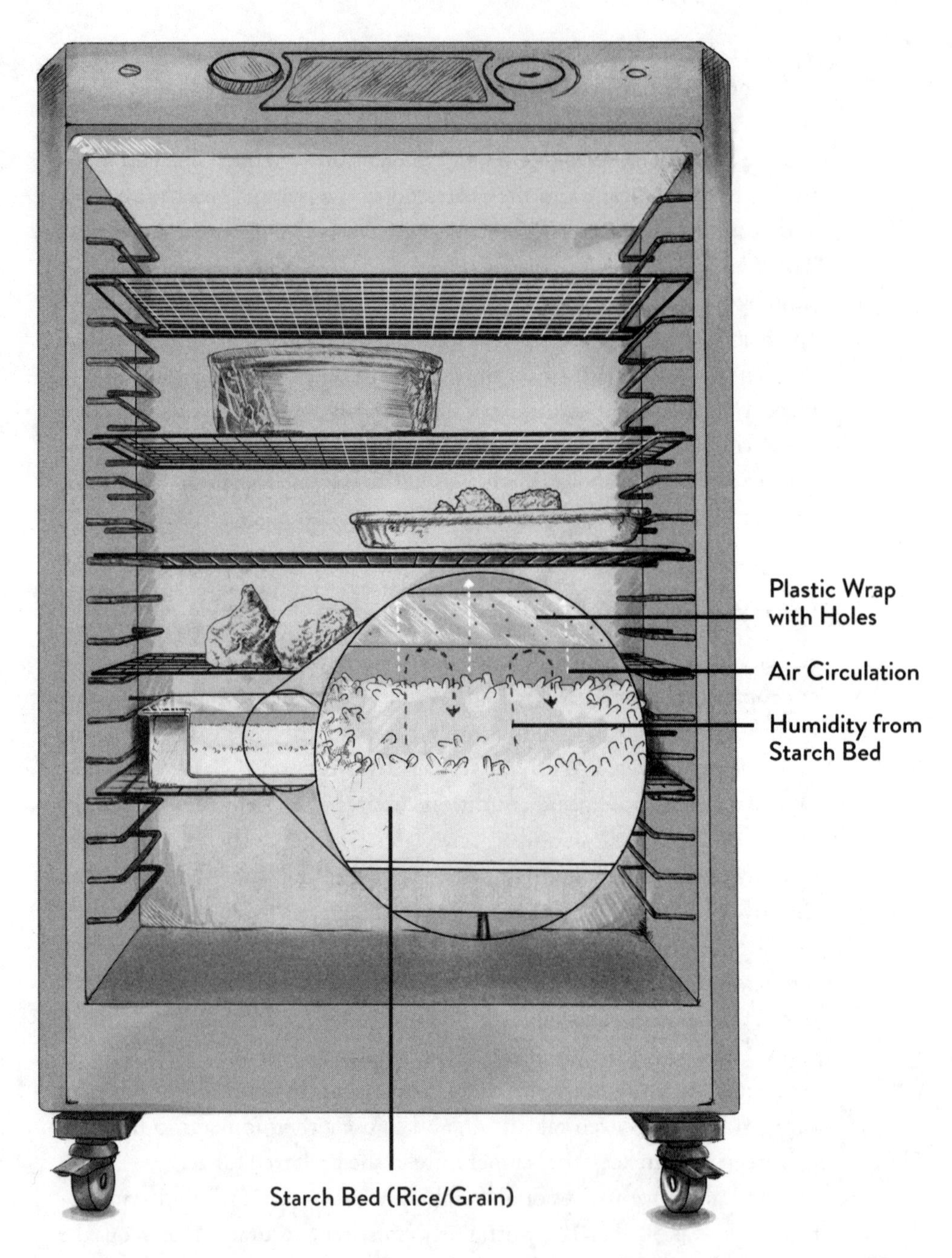


Dehydrator incubator details. *Illustration by Max Hull.*

## Bare-Bones Oven Method

If you aren't sure if you want to invest anything in this process yet, you can use what's available in your home. The oven in your kitchen is an insulated cabinet designed to hold food at temperature. For some ovens, leaving the light on inside is enough to generate the heat required. Some gas ovens have a pilot light that burns continually and can be a heat source. In this case, the door closed may be too warm for koji, but simply cracking it ajar is enough to get to the desired temperature range. Keep your trays toward the bottom of the oven where it's cooler to avoid possible overheating. For any of these methods, you'll have to measure the temperature of the inside of the oven with an accurate probe thermometer. The overall message here is that it's not difficult to hack something together with any already available heat source that is slightly over room temperature to grow koji. You can even do it whenever the weather conditions are right.

## Bread Proofer and Other Commercial Kitchen Options

As we have discussed, commercial bread proofers are another option for providing favorable conditions to feed yeast. These are especially useful for consistent results at large production quantities. What's wonderful about bread proofers is that they can be set to the lowest optimal temperature for koji at 86°F (30°C) as well as offer the humidity required. These systems can hold multiple full sheet pans (26 × 18 inches / 66 × 46 cm) and range from around 2½ to 5½ feet (0.8–1.7 m) in height. That's a lot of koji making potential!

Our friend Sam Jett suggests using an old refrigeration unit as a modern muro (insulated cabinet). All you have to do is incorporate the controlled humidifier and heater system described in the "Eric Edgin's Muro Humidifier and Heater Method" sidebar.

Most serious restaurant kitchens have specialized combination ovens that can accurately maintain temperature and humidity to grow koji. Just set it and forget it. The only disadvantage of these amazing ovens is that they are heavily used; sequestering them for forty-eight hours straight is difficult.

## ERIC EDGIN'S MURO HUMIDIFIER AND HEATER METHOD

One of the best modern versions of a traditional Japanese *muro* (incubation chamber for maintaining a proper koji environment) is handcrafted by our friend Eric Edgin. He is the one of the only people we know who has made custom, stand-alone systems out of cedar. The design is well thought out as a result of his research and years of koji making experience.

The muro is a cedar cabinet that is approximately the size of a small oven. It has shelves to hold the koji trays at the top. The bottom has a personal heater as a heat source and small room humidifier plugged into standard temperature and humidity controllers set to the appropriate conditions. The following description is adapted from Edgin's former website:

*The incubator is made of cedar. It is the ideal wood for a muro for a number of reasons. One, cedar can be used untreated, which allows for it to absorb any moisture that would condense in the muro. The cedar is naturally rot resistant so can take the moisture without compromising the longevity of the cabinet. The cedar can stand up well to the heat and humidity changes from the cycles of use and disuse. The cedar will patina well over time.*

*The cabinet is built using mostly mortise-and-tenon as well as dovetail construction. The lack of iron (potentially corroding) fasteners and hardware inside the muro will help it last a lifetime. There are gaps built in for allowing the expansion of wood with the increase of humidity when the muro is in use.*

---

Growing koji is a fun and rewarding process, and every time you do it, you are acting as a shepherd of a nine-thousand-year-old food tradition. The magnificent thing about this tradition is that it is alive and well, growing and maturing in many ways. When you grow koji, you are in control of how you want to produce it and what you want to make with it. This act of instilling a bit of yourself into the process is what makes food, cooking, and cuisine so intimately special. You can weave the story of who you are into a very tangible sensual experience that can be shared with others. Grow forward!

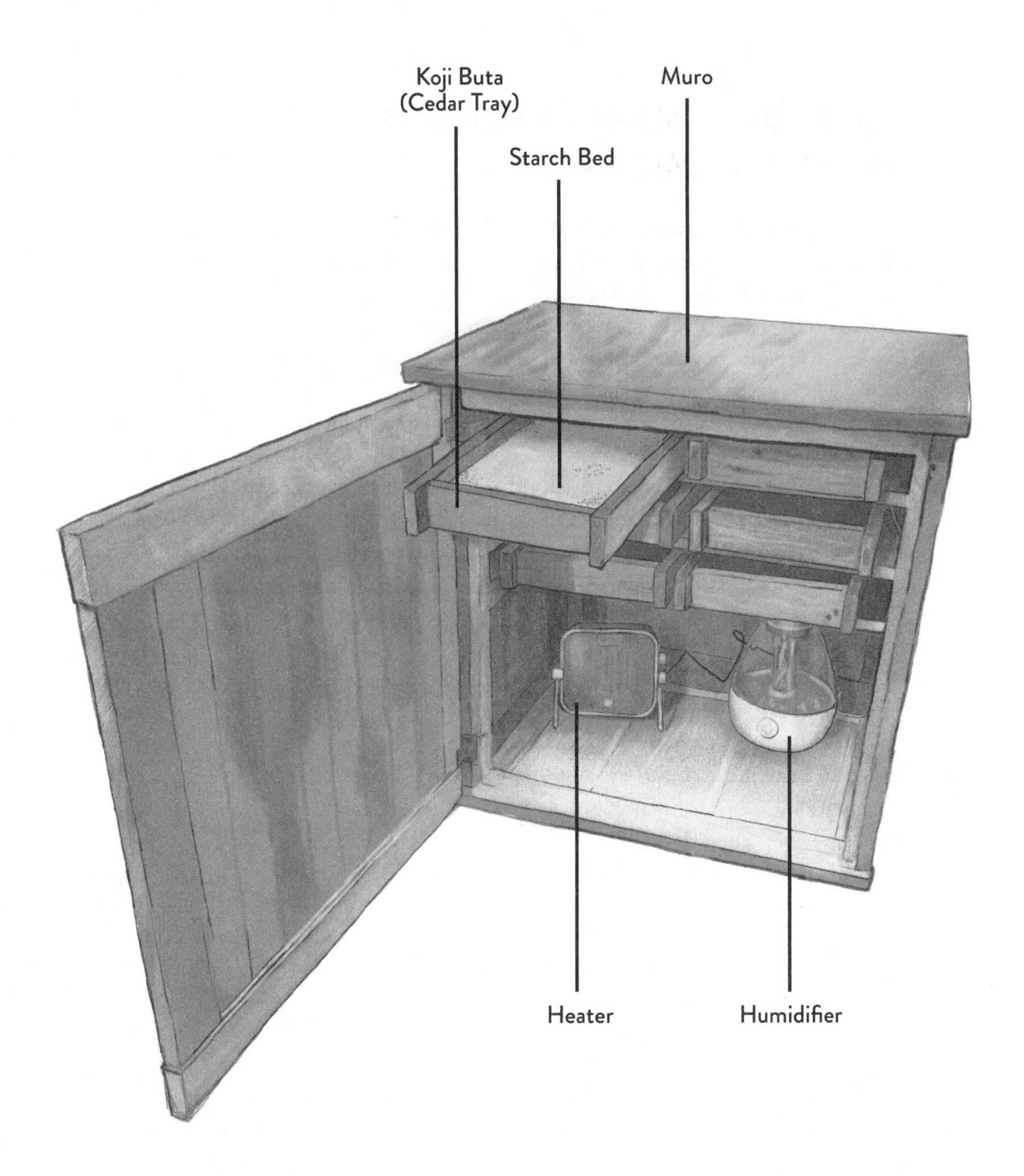


Humidifier and heater incubator details. *Illustration by Max Hull of Eric Edgin's muro.*

## THE FERMENTABOT

### from an interview with Eugene Zeleny of the MIT Media Lab

The ability to track and control small-batch fermentations has not been historically available for small kitchens, nor have there been industry-accepted standards to produce the variety in micro-batches. Now a collaboration between the Basque Culinary Center and MIT Media Lab's Open Agriculture Initiative, the Fermentabot, hopes to combine the worlds of food upcycling and fermentation with engineering, control systems, and data collection.

The Fermentabot minutely adjusts temperature and humidity levels (as well as pH, oxygen, and carbon dioxide in future iterations), monitors how ingredients react to specific environments, and adjusts the controls based on a pre-programmed recipe. Small variations in temperature and humidity can drastically affect enzyme levels during fermentation, and gathering data on the process proves incredibly useful to make food production more reliable, consistent, and resource-efficient without sacrificing flavor. Imagine reusing your kitchen's food scraps, such as carrot peels and the outer leaves of a head of cabbage, to produce more food at the touch of a button. This is one of the main intentions for the Fermentabot. The Fermentabot can potentially have a large impact on the way individuals make food. So far, small-scale fermentation has been a mostly manual process, with no documented engineering efforts involved in improving the efficiency, output, or factors that affect spore growth. The Fermentabot allows monitoring of the fermentation process, and has the potential to provide feedback to producers, chefs, and educators, both professional and recreational.

This project has been lightheartedly called a "very fancy sort of oven" by Diego Prado, formerly an associate professor at Basque Culinary Center, and he isn't too far off the mark. The Fermentabot was designed to utilize as many off-the-shelf components as possible, including heaters, fans, humidifiers, nuts, bolts, and catering trays, all attached and integrated into an insulated food pan carrier. Mechanical, electrical, and software engineers designed the housings, fluidic systems, printed circuit boards, and wiring. The custom-designed components were laser-cut and assembled in Open Ag's fabrication facility in Middleton, Massachusetts. While not everyone may have access to laser cutters or makerspaces, there are ample opportunities for the fermentation-loving communities to innovate and find easier and cheaper ways of building their own personalized Fermentabots.

Similar to the rest of the Open Agriculture Initiative's projects, the Fermentabot design, build instructions, bill of materials, and collected data are all open source and available online for free. Instead of buying fermented products, people can make them at home, thus redemocratizing fermentation. The hope is for the Fermentabot to become a standardized and commonly used tool, the trickle-down effects poised to be impactful.

Down the road, transportation costs can be reduced, since the ingredients can be sourced locally. Non-technical individuals can use climate recipes uploaded by other Fermentabot users and can create new recipes by using the web-based user interface. This way, identical koji batches can be grown and reproduced all over the world. Their prediction is that this is just the beginning of what's to come, and there's no telling what sort of innovations the worldwide fermentation-loving communities will think up with this capability.

## BASIC RICE KOJI MAKING QUICK REFERENCE

1. Soak rice for at least 8 hours, drain, and steam until al dente.
2. Place rice in a hotel/ baking pan or tray and allow it to cool to body temperature.
3. Sprinkle koji starter over the rice and mix well. (1 g starter per 1000 g rice)
4. Spread flat without compressing the bed at a height of 1 to 1.5 inches (2.5–4 cm).
5. Cover the tray of rice as prescribed by the incubation setup for the remaining steps.
6. Put the tray in the chamber and incubate at 86°F (30°C) for 12 hours.
7. Remove the tray, mix the rice, spread flat, and incubate for 12 hours.
8. Remove the tray, mix and mound the rice then incubate for 12 hours.
9. If the rice is bound with mycelium, the koji is ready. If not, repeat step 8 once more.

CHAPTER 5

# Expanding Your Koji Making

The focus of this chapter is to dive deep into making different koji types beyond a simple white rice preparation. You'll find guidelines to optimize conditions for *Aspergillus oryzae* so you can use almost any starch. We will cover how to optimize the cooking process based on major grain types. We have also included a discussion on guidelines for preparing any non-traditional carbohydrate medium, such as popcorn, for successful growth. Finally, we'll review the different types of spores and the information required to decide which strain to choose.

When you make koji, it's important to understand the right conditions, from nutrition to the environment, to be successful. At this point you understand the basics enough to realize that making rice koji is quite simple. We are now presenting what will allow you to make just about any koji you can think of. We will walk through the four key factors—creating a microbially clean slate, maintaining temp, maintaining humidity, and creating an accessible starch with structure—to help you make informed decisions about how to optimize your process to achieve good and repeatable results.

## Create a Microbially Clean Slate

One of the most important factors to ensure food safety when making koji is establishing growing conditions with no harmful microbes. This is no different from pretty much every other controlled fermentation process that isolates a specific culture and feeds it a known medium to yield a specific result. Remember that *Aspergillus oryzae* generates no mycotoxins, unlike most other *Aspergillus* strains, because they have been bred out. It has survived as a result of humankind nurturing it for thousands of years, originally for survival and today, for use in preparing delicious food. The mold feasts on a nutrition-rich environment without competition because everything is cooked *prior* to its application. Grains or legumes are steamed, boiled, or pressure-cooked; grains or flours are toasted. Of course, the circulating air and contact as a result of mixing brings in microbes that come along for the ride. However, since the mold spores are introduced as soon as the environment is welcome, they grow rapidly under ideal conditions and, as long as no cross contamination has occurred, will outcompete others that may attempt to take hold.

## Maintaining Temperature

As we have discussed, *Aspergillus oryzae* requires a specific temperature range to grow favorably. However, it's not as easy as setting your incubation chamber to a hold temperature without considering the exothermic reaction when a mold starts to grow. The chemical reaction generates heat in the medium. In traditional koji making, heat is managed by a mixing schedule to fluff the grains and dissipate heat. After mixing, the top of the grain bed is furrowed (mounded in rows) to create additional surface area to help with cooling. The height of the grain bed is also an important factor; a too-tall bed would result in the grain packing together too densely, making it more difficult for the heat to dissipate. Somewhere between 1 and 1½ inches (2.5–4 cm) in height is what we recommend.

Preventing overheating is important because when the grain bed overheats, the *A. oryzae* gets stressed and sporulates (goes to spore/seed) to protect itself. Ultimately, this halts enzyme production, which impacts the koji's efficiency when you create various koji-based makes. In our experience, overheated koji is still usable. You just need to know what to look for and be aware of what is compromised. In general, if you get to the thirty-six-hour mark with an overheated medium with obvious mycelial growth, you still have plenty of enzymes to work with.

The key issues of overheating are sporulation and natural fermentation. If the starch bed remains overheated long enough, the blanket of spores you find on the top of your grains will create an off-flavor when used in short-term applications (such as marinating a cut of meat in koji or adding koji to a cocktail). You also may find that the bed has started to naturally ferment as a result of the microbes that come along for the ride in the air—most often, naturally occurring *Lactobacillus*, found in everything from sauerkraut to yogurt, which creates a slight souring in the medium. But this is not necessarily a bad thing when it comes to most applications, as a touch of acidity can often balance a flavor profile.

Another important consideration in temperature management is how it relates to enzyme production. Koji makes amylase in the temperature range of 95 to 105°F (35–40°C), but too much of this enzyme production yields sticky koji. To mitigate this issue, you can cool it by mixing it between the eighteenth to twenty-seventh hours of incubation. Ventilation can be helpful, but overcooling can bring the process to a halt. Fungal growth slows at around thirty-six hours and start building important proteases below 95°F (35°C).[1]

In summary, warmer conditions are better for amylase production, and cooler conditions for protease to convert proteins into amino acids. Your final application will determine whether you lean toward one or the other, whether the sweetness of an amazake or the umami of a soy sauce. Use this dialed-in temperature consideration if you have the control in place and want to optimize it as such. However, we've found that setting the temperature on the lower side, between 86 and 95°F (30–35°C), works quite well and generates enough enzymes to be used in practically any application.

## Maintaining Humidity

Humidity is another environmental condition that is important to growing koji. A damp environment maintains the structure of the gelatinized starch grain, making it easy for *Aspergillus oryzae* to consume. We've found, however, that maintaining a strict humidity percentage is not necessary in the setups that we've used. We use the naturally occurring moisture in the cooked rice, grain, or other medium in a container combined with limited ventilation, as well as using a fully contained water bath incubator (with water condensing on surfaces from constant humidity generation). The only challenge with the water bath incubator or any constantly producing humidifier is managing the condensed water that can drip onto the starch bed. Koji will not grow well—or sometimes not at all—on an overly wet mushy surface. A too-wet, surface may also allow other undesirable microbes to proliferate. Fortunately, it's easy to manage the condensation by placing a damp lint-free towel on the grain bed. Overall, you don't need to worry much about humidity when it comes to using the incubation options we've presented.

## Create an Accessible Starch with Structure

What does accessible starch with structure mean? It's probably easier to explain what it isn't. It is not an uncooked grain with the bran intact, such as hulled wheat berries. Koji cannot grow on their cellulose-covered kernels, and their uncooked internal endosperm is too tough for it to penetrate. However, an accessible starch with structure is also not porridge, as that would be too wet for the mold to grow—the starch requires a specific moisture content. To understand the proper preparation, let's begin with tried-and-true processes that have been around for centuries.

The four preparations that will give you a firm grasp of the basics are: polished grain (best known as a key component in miso making and the standard for driving a lot of the makes in this book), soy sauce mash (a mix of cooked soybeans and toasted, cracked wheat), *douchi* (known as Chinese fermented black beans with a base of cooked soybeans coated with toasted flour), and *meju* (a mash of soybeans that's formed into a block and hung to be fermented). Each of these preparations, detailed below, will give you a foundation of knowledge for how to create your own koji-friendly starch beds beyond just using white rice. Extending beyond these fundamentals, we've developed other starch preparations to grow koji. We hope that the gamut of traditional and modern ways to create a koji-friendly food source described in this section will help you find your way into koji making. As you can see, there are many ways to achieve it, some more challenging than others. The idea behind all of these derivations is to provide as many "hows" as possible to make koji with *any* starch. Try the method that speaks to you. This fundamental understanding will ultimately allow you to formulate your own process using whatever resources you have available. It's about building intuition . . . but we all do it in different ways.

**Polished Grain Koji.** Preparing koji with polished, bran-removed white rice is by far the easiest process to follow and accomplish, as we've already discussed (see "Making Basic Rice Koji the Traditional Way" on page 64). The reason we call for an "al dente" cook is because the starch becomes gelatinized while the grains retain their shape. This is ideal for creating an easily accessible food source for the mold to consume and infiltrate. Because the individual grains maintain their shape, there is plenty of surface area that mycelium can extend into and branch out from.

**Soy Sauce Mash.** The koji base for soy sauce is equal parts cooked soybeans and cracked roasted wheat.[2] Soybeans are much lower in carbohydrates and higher in protein than cereal grains. It is challenging for koji to grow solely on soybeans since they don't offer nearly as much nutrition to feed on, and their surface structure tends to be too wet. In hybrid mash, however, the cracked wheat boosts the starch content and soaks up excess water so the medium is much less wet and sticky. Each wheat kernel is broken into four to six pieces, which offers plenty of exposed surface area for available carbohydrates. The neat thing about this combination is that the high-protein medium is already incorporated, so it doesn't need to be added later as it does when you're making miso. The toasted grains also add flavor complexity.

**Douchi: Chinese Fermented Black Beans.** Douchi can simply be defined as fermented black soybean koji. You may have tasted douchi in the umami salty bits in the sauce of pork ribs, or the clams served in tiny bamboo steamer baskets from a dim sum cart. Functionally, douchi is a whole soybean miso, and the koji is pretty much made in the same manner. However, fermenting individual beans creates a very different, cravable flavor. Similar products are made in Japan (*hamanatto*), the Philippines (*tao-si*), and even India (*tao-tjo*). The methodology and actual species or variant of mold for making each one differers from locale to locale.

The preparation is the same: The beans are soaked, then cooked (steamed or boiled), then cooled. Wheat flour, somewhere between 1 and 6 percent of the somewhat dried, cooked soybeans, is toasted and mixed in with the mold spores to inoculate the beans.[3] The readily available starches in the wheat coating the beans help get the koji started. They also absorb water to keep the wetness at bay and make it conducive for growth. After that the beans are rinsed to remove the koji spores, which can impart a bitter taste in the finished product. The beans are then brined, sometimes with the addition of chilies, garlic, and ginger; then dried until they have a soft, almost fudgy texture.

Another approach to producing douchi is to culture the beans directly with any of the available strains of *Aspergillus sojae*. We find this method to be very efficient, as it doesn't require the addition of wheat flour or another starchy substrate to allow for growth. Some varieties of *A. sojae* (such as Sojae No. 12 from Higuchi Matsunosuke Shoten Co. Ltd.) create amazingly fruity aromas when growing that are reminiscent of mango and pineapple. This aroma ends up carrying over as an underlying note in most koji makes that use it, including douchi that has been brine-fermented and then dried.

A variable that we find fascinating is dry salting the douchi instead of brining it. We salt the beans to 3 percent should we want a lactic tang to develop or 7 percent if we want a more koji-forward flavor and heavy umami taste. At either percentage we allow the beans to ferment or cure for several weeks before rinsing and then drying them. Many people have commented that this dry-salted douchi reminds them of the Mexican sauce mole. The beans seem to develop aromas and flavors that are chocolaty and raisiny. They work well in many Mexican dishes and Eastern European ones, such as cholent or tzimmes.

**Meju: Korean Soybean Starter.** Meju is a dried fermented soybean block that is the base flavor component for Korean condiments, the most familiar of which these days is *gochujang*, a spicy chili paste commonly used in marinades, sauces, and soups. The basic process for making meju is to cook soybeans to make a thick paste that can be formed into blocks or balls, then partially dry them until firm and hang to allow them to naturally culture. Traditionally, rice stalks are used to surround, support, and tie up the bricks. As a result the naturally occurring microbes from the plant and in the air "seed" the meju. Over generations of making in specific environments, fermentation structures, and preparations, each maker develops a unique flavor driven by terroir. In commercial making, however, the medium can be inoculated with a known mix of microbes to assure consistency. Culturing meju is very different from culturing the other three preparations discussed in this section, as it is driven by a variety of microbes instead of a primary isolated one. As a result of the diversity of microbial activity, it takes longer for the final product to fully develop. This process is much closer to aging meat and cheese than growing using other techniques we just described. However, given the multitude of microbes in action, starting with a crazy complex base for a secondary fermentation has its advantages.

If you're wondering about meju's more natural fermentation environment with regard to the potential for unwanted microbes to take hold, we would argue that the process to produce naturally fermented meju has been upheld for centuries without any problems. The more we investigate and explore traditional koji and fermentation techniques, the more we see that following established practices of preparation and cleanliness yields a safe product.

## Tips for Moving Beyond Tradition

As much as we love and respect these four traditional processes, we like to break them down to their cores and see what makes them tick, then adapt them to our more modern times. Do we need to follow every single step exactly? What are the fundamentals that make them work? Why can't we do something another way? Was the original method driven by availability of ingredients, equipment, technology, or the like? Can we leverage our modern knowledge to get to a comparable end point more efficiently? Armed with a basic understanding of traditional, well-established processes to culture koji, we developed a series of ideas to make mold proliferate in all sorts of starch-rich environments. This is what we've found works for us and many friends from around the world with whom we've shared this information. Just remember that nothing here is set in stone or necessarily better than other methods out there. This is just our selection of options to consider—you can determine what's best for your needs.

## GRAIN SIZE AND STRUCTURE MATTER

Once we had a grasp of the traditional method of making rice koji, we started thinking a tiny step beyond to optimize the process. Rice koji is typically made with medium-grain rice. We wondered: Why not long-grain? The thinner midsection of the grain would allow for faster hydration, cooking, and mold infiltration to the core when applied to the standard process, which ultimately yields a more efficient koji production process. Also, long-grain rice has the highest percentage of amylose, which koji is quite fond of as a food source. We noted that this process yields a nice product at the thirty-six-hour mark, when the standard rice koji with medium-grain takes on the order of forty-eight hours. Extending slightly beyond this idea, we wondered about a product called *broken rice*, which is exactly as it is described. Most large Asian markets carry broken rice, the imperfect broken bits of rice left over as a result of processing. In Thai cuisine it's used to make a jasmine rice porridge. The beautiful thing about this is that it's perfect for koji making and much less expensive.

One of the biggest challenges with making koji with grains other than white rice (or other polished grains like pearled barley) is that their bran is still intact. In the cases of brown rice, wheat berries, barley, and rye, for example, the traditional soaking and steaming process to prepare the grain isn't enough to split the bran sufficiently for the mold to infiltrate. A great way to overcome this is to break up the grains into halves and quarters after steaming, similar to the technique of cracking wheat into smaller bits to assist in inoculation in a soy sauce koji mash. However, we recommend not breaking up the grain until *after* it's been cooked. We've found that cooking grains that are already broken up tends to make them more wet and sticky, which is not optimal for growing koji. Also, it's easier on most food processors to break up the softened grains. Some home food processors don't have a fast enough blade speed to cut the steamed grains, so you may have to resort to hand mashing or pulsing small batches in a blender. Another option is putting it through a meat grinder. Just make sure it's sanitized first. Regardless of the machine used, be sure not to break up the grain too

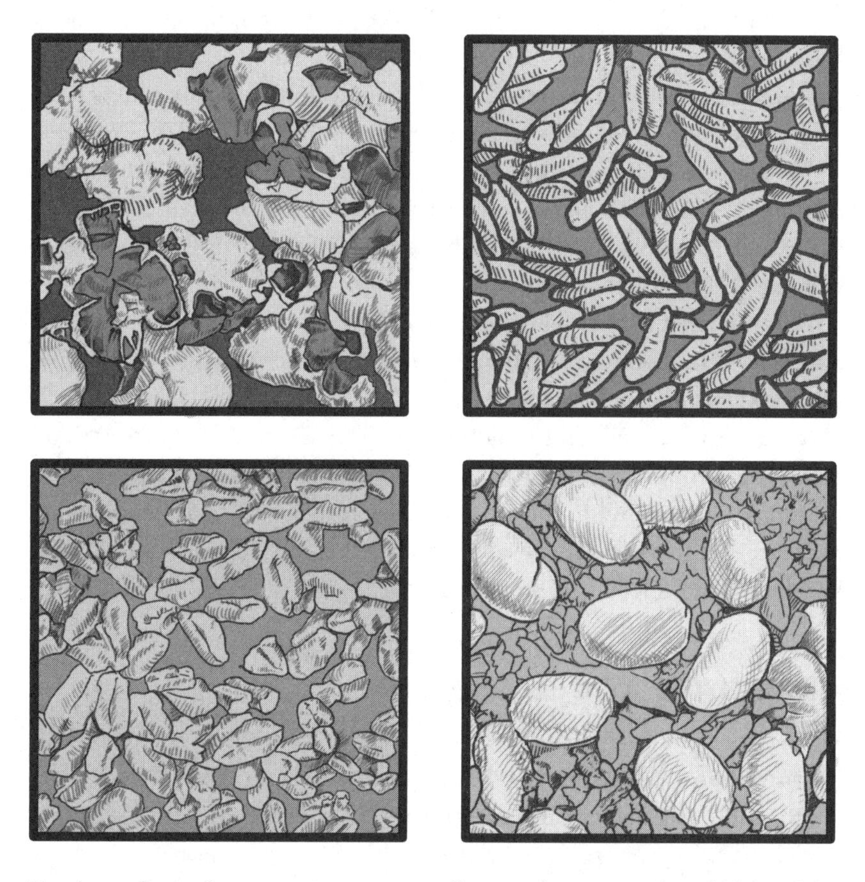

Koji bases for surface-area comparison. All are to the same scale. *Clockwise from the top left*: wetted popcorn, jasmine rice, soybeans and cracked wheat, and rough broken-up wheat grains. *Illustration by Max Hull.*

much, creating unfavorable wet conditions. You need those important air pockets within the grain bed for an optimal koji growing environment. There's a balance to be struck.

Another way to make the starch accessible is to cook grains by boiling them in water until they blow out of their bran coat. This technique creates the most accessible starches and holds on to the bran, resulting in an individual grain structure conducive to creating air gaps in the grain bed. However, you may be asking why we're now recommending an overcooked starch that's wet and not conducive to growing koji. Well, one addition to the process makes this work. Toasting and grinding a flour using the same grain and then dusting the boiled mash with it allows you to manage the moisture appropriately. As you may recall, coating with flour is nothing new—it's used in the douchi making process to help the mold take hold on soybeans.

When making masa (also known as hominy) from dried corn, a cooking process called *nixtamalization* is used to remove the seed coat. Once the kernels, post-cook, are mostly broken up into quarters, it becomes a wonderful koji growing medium.

One quick starch cooking process we have tried doesn't require water: popcorn. When you apply enough heat to the kernels, the pressure of the cooking starch bursts the seed coat and puffs it. The aerated starch becomes very accessible for mold growth. This is one of the first entirely off-the-beaten-path methods that we tried, and it's by far the fastest grain preparation. The only problem is that popcorn doesn't have very much moisture, but this is easily solved by misting the popped grains with water. See the "Popcorn Koji" recipe on page 107.

## ADEQUATE ACCESS TO AIR

Filamentous mold requires oxygen to grow. We have said it before, but we can't say it enough: Maximizing the surface area of starch exposed to air is the key to yielding a good koji. In a traditional polished grain bed, the individual kernels are piled on top of one another, which naturally creates an internal network of air pockets. That's why it's important that each grain has enough structure to be separated as much as possible from the others. Any surface area that is in contact with an adjacent grain does not have access to air.

The softer the grain, the greater the likelihood of more contact area and less air access. The structural integrity of the grain is important when it comes to how high the bed is—the layers of grains at the bottom will get more compressed and squished together, with less opportunity for access to air pockets. The shape of the grain is also a factor in how the pieces stack to create gaps for the mold to grow. We've seen folks using perforated hotel pans lined with kitchen towels to allow for additional access to air. This is a technique that the folks at Noma employ when they make koji. We've also seen Shola Olunloyo of Studio Kitchen expand upon this idea with the use of wire-mesh baskets for even more air exposure.

### 3-D PRINTING

What if you could design any shape to create the perfect conditions for an accessible starch? Imagine a pastry bag with a fine tip attached to 3-D printer head. Now imagine an optimal gelatinized starch being extruded to create a 3-D web networked together that would have the ideal thickness for mycelium penetration at maturity, as well as just the right air space for circulation and heat dissipation so mixing is not required. This is a project that we're looking for someone to work with us on, as we feel it has potential to optimize koji making efficiency even further.

## Diving Deeper with Koji Spores

There are many different types of koji. If you're not planning to use our universal light rice method, you'll need to know what kind of spores you want (and then source them) before you even think about what setup you'll need to grow koji. The first step to knowing what kind of spores to source is having a clear vision for what you want to produce. Are you a brewer who wants to incorporate koji into your malting process? Are you a chef or butcher who wants to use koji to create charcuterie or age-accelerated meat? Are you a baker who wants to incorporate it into the hydration for your breads? There are also dairy-based applications for cheesemakers and, of course, many choices for those interested in making amino sauces and pastes. Production volume should also be a consideration. Are you making

## THE IMPORTANCE OF PH

### Sam Jett

*One factor that the two of us haven't spent a lot of time on is how pH influences koji growth. In all of our experimentation, we haven't had much trouble getting koji to grow on most everything, so our focus has been elsewhere. However, a couple of years ago, our friend chef Sam Jett, co-owner of Patchwork Productions with chef Sean Brock, was posting about how he dropped the pH of cocoa nibs so that koji would take to it. We couldn't think of anyone else better to present the subject of pH.*

Koji likes its environment humid, but not too humid. It likes to be warm, but not too warm. The same goes for pH. Surprisingly, koji is pretty pH-tolerant compared with other fungi. Through a lot of tests on growing koji on different substrates, I found that many of their natures can be adapted to help the culture be successful. If the plan is to make cocoa nib koji, for example, then understanding that cocoa nibs are *too acidic* is important. Even the pH of your water influences the substrate's composition. If you utilize towels to cover your koji during inoculation, were they treated with bleach? Having a firm understanding of these details and being mindful of the steps in this process will only be beneficial to production.

Koji likes a neutral environment. *Aspergillus oryzae* tends to grow the best for me when I keep the pH of the substrate around 6 to 7.8. Anything under 5 or over 8.3 has resulted in the death of the culture. The good news is this gives us plenty of wiggle room—most foods naturally fall within this range, and the foods that don't are easy to modify. After hominy has been treated, for example, it requires a lot of washing and boiling to drop its pH to a neutral enough state for consumption. In the example of the cocoa nibs, their pH of 5 can be brought up simply by boiling them for a few minutes, essentially hydrating various starches that the koji wants to consume. Another way to adjust pH is to utilize pH buffers such as sodium citrate or different types of powdered acids. Just create a solution and cryo-boil or completely vac the substrate with it.

The area in which we do our inoculation is just as important to understand. When I wash my utensils, I do so with very strong sterilizing chemicals. Multi-quat sanitizers are meant to be left on and dry to help prevent organic growth. If I treat my cover towels with bleach, surely some will leach out to my culture during inoculation. I am not suggesting being unsanitary (having safe practices is the most important part of fermentation), but knowing about each aspect of your overall environment is important.

Once the koji is made, it's time to ferment. Koji tends to have increased amylase and protease production in more acidic environments. The most successful trials I have had in fermentation range around 5.3 to 6.5 pH. Anything lower than 3 or above 8.5 would result in the reduced enzymatic activity. This gives us a wide range of ingredients we can use and a wide range of products we can create! Hominy miso, check. Buttermilk shoyu, check. Banana miso, check. Apple shio koji, check. Country ham "garum," check. The sky is the limit. It is all about trials and finding a process that works for you. Once you have that, practice and dial in the procedure. Then practice again. It will always be about the process.

5 gallons (20 L) of a product or 100 gallons (380 L)? Through several years of practical work and testing, we've narrowed down specific species and strains for specific purposes. We'll be going into these details later on in this book.

It's important to note that there are not only several species of *Aspergillus* used in food production but also countless variants of each species. Each of the koji producers in Japan has bred its own variants and named them, much like the producers of vegetable seeds for their different types of vegetables. These variants differ from one another in that they have each been bred through selective breeding to optimize a different attribute, which includes the optimization of specific enzymes, specific aroma creation, and even the production of citric acid. We offer in this section a listing of some of the different species and variants that we work with. The listing is a selection from the product line that Koichi Higuchi's company, Higuchi Matsunosuke Shoten Co. Ltd., offers. It is broken down by their intended use such as the production of amino pastes, amino sauces, and various alcohols. As this book progresses you'll see that the uses outlined below are suggested and not hard and fast. Any of these species and variants can be grown on virtually any substrate with enough care and used in myriad creative and exciting ways.

### SAKE

**Hikami Ginjo** (*Aspergillus oryzae*). This variant is resistant to the high temperatures reached during the Ginjo koji making process and produces a koji strong in amylase yet weak in protease. Its strong amylase action is perfect for making alcohol and vinegar.

**Kaori** (*A. oryzae*). This new type of koji fungus creates incredibly aromatic and fragrant koji. *Kaori* translates as "fragrance." This is a favorite of ours for producing highly aromatic alcohols and rice vinegar.

### SHOCHU

**Brown Koji Fungus for Barley** (*A. luchuensis*). This variant is characterized by exceptionally high levels of citric acid and enzymes. When used with rice instead of barley, the enzyme generation is exceptionally high but the citric acid production is somewhat lower.

**Black Koji Fungus** (*A. luchuensis* var. *awamori*). Black koji fungus, an ancient koji, is back in fashion after being neglected for many years. If you want a shochu with a unique bouquet, this is for you. When used with rice the citric acid production is on par with brown koji fungus, but when used with barley it is inferior. We love using this koji to ferment cream cheese and other fresh cheeses.

### MISO (AMINO PASTE)

**BF1 for Rice Miso** (*A. oryzae*). This can be used to make pale-colored types of miso and amazake. It has strong saccharifying power and the strongest protease action. This is our favorite all-purpose spore; we use it for all manner of foods, from amino pastes to amino sauces to amazake.

**BF2 for Rice Miso** (*A. oryzae*). This can be used for pale and for red types of miso. It has the strongest saccharifying power and strong protease action.

**Barley Yellow Koji for Barley Miso** (*A. oryzae*). This can be used for red and sweet types of miso. It has strong action for both its saccharifying power and protease development. We love the strong earthy flavors that this variant pulls out of barley and other earthy grains such as rye.

**Hashimoto for Soybean Miso** (*A. oryzae*). This is known for having a good fragrance that is reminiscent of mango. It has medium saccharifying power and strong protease action. This variant cultures beautifully on every legume we've encountered. It becomes extremely fragrant on white and great northern beans.

**White Moyashi for Rice Miso** (*A. oryzae*). This is a variant that has white spores when mature and is a favorite of ours for amazake and shio koji. It is also great to use if you want to avoid off-colors and -flavors caused when your koji goes to spore.

### SOY SAUCE (SHOYU)

**Sojae No. 9** (*A. sojae*). This is ideal for when a pale sauce is desired. The most distinctive traits are the highest protease generation available and minimal dust due to the low level of koji spores, thereby providing an excellent sojae fungus. This also works great for charcuterie and other applications using ingredients with large amounts of protein in them.

**Sojae No. 12** (*A. sojae*). This fungus strain brings out the characteristics of sojae most clearly. It is most suitable in situations where sojae's unique fragrance and a pale raw sauce are required. We love the fruity fragrance of this koji, which is reminiscent of mango and pineapple. It's a favorite of ours for making amino pastes solely out of beans and for making douchi.

**Hi-Sojae** (*A. sojae* and *A. oryzae*). A koji starter that is a fungal compound of mainly sojae fungus with *A. oryzae*. We recommend this koji starter when a koji providing the characteristics of sojae fungus but with low fragrance levels is required.

## A FAMILY BUSINESS

On our journey down the rabbit hole of koji adventures, it only made sense to expand into the wider range of possibilities by finding a direct source for our spores instead of sourcing from a spore importer in the US. At first we were a little nervous about whether or not a traditional koji starter maker would accept our modern experimentation. To our surprise, we found a gentleman who welcomed us with open arms when we told him about what we were doing with koji. We became fast friends with seventh-generation maker Koichi Higuchi, of Higuchi Matsunosuke Shoten Co. Ltd. in Osaka, Japan. He is intrigued by all the recent developments in applications around the world and is looking forward to seeing the changing landscape seeded by his traditional products.

### ORIGINS OF KOJI STARTER

The origins of koji starter are unclear, as there is no documentation. Basically, it is supposed that, like plant seeds, koji was cultivated over a long period to isolate the desired spore performance. Once the optimal characteristics were realized, those specific spores were used to seed the next batch of koji. The problem with this method was that after a few cycles of growing, unwanted bacteria contaminated the quality of the koji and resulted in bad sake. To maintain quality, wood ash was introduced to suppress contamination. It was also found that wood ash contained minute elements of phosphorus and potassium, which acted as a preservative. From the 1900s on, sterilization techniques improved and wood ash was no longer used. Unadulterated cultivation of microorganisms became established.

When Higuchi's family began making koji starters, they only had two different types, one for the sake starter and one for the *moromi* (sake mash). These originated from three strains. Since then, the company has grown to offer fifty different types of koji fungus strains used for flavor and focused on enzyme productivity in all of the traditional Japanese fermented foods and drinks: amazake, sake, miso, shoyu, shochu, mirin, and vinegar. Also, for product diversity, they combine strains to leverage the benefits from the mix. For example, a koji starter used in the production of sake will have a combination of types to enhance and balance the complexity in the delicate taste and aroma.

It all started in Osaka, a thriving trading port city, where Higuchi's ancestors were merchants. As a result of that history, a lot of large companies became established there. The city is conveniently positioned between two large sake manufacturers, Nadagogo in Hyogo and Fushimi in

Kyoto. Fortunately, the Higuchi family's koji company was strategically close to both in an era when distribution was difficult. Also, being in the vicinity of a number of sake makers in the area today allows them to quickly react to feedback on their products and make the necessary improvements for future production runs.

From the beginning, their aim was to enable the stable production of sake. The main goal has always been to produce a koji fungus product that is as unaffected by bacteria as possible. After World War II, in the midst of material shortages, they endeavored to create strains with high enzyme activity to maximize sake productivity. Further progress in product development for enzymes and sake was made in the 1970s. There were two problems in sake production: black lees (left after pressing) caused by tyrosinase, and the yellow discoloration of sake caused by a reaction with iron. Both were overcome by developing koji fungi that did not produce these elements. From the 1980s on they have developed koji starter to cater to elegant, high-quality sake and more recently the Ginjo sake boom.

This market focus yielded a koji starter Higuchi is really excited about—a customized formula for Ginjo sake, considered the pinnacle of sake. Sake brewers who make Ginjo sake have very strong opinions. The base koji is paramount in this sake—the rice is polished to a high degree, and temperature and humidity have to be closely monitored constantly during the koji making process. Veterans of Ginjo sake making highly value the production of excellent koji. Based on an analysis of koji from all over Japan, the Higuchi family has accumulated several hundred strains suitable for Ginjo sake, and from these they decide on a well thought-out combination to make Higuchi's product. Over the last twenty-five years of production, the producers of Ginjo sake who use Higuchi's starters have been enamored with the results. We see this as an amazing marriage of technology and craftsmanship.

### MODERNIZATION OF KOJI

As a koji starter maker, Higuchi Matsunosuke Shoten Co. Ltd. has moved away from traditional methods. The primary reasons are continuous humidification causing contamination issues and unsuitability for producing a wide variety of koji fungi. They are now using science laboratory technology for their production process, which maintains the traditional intent to make a consistently pure product. *Aspergillus oryzae* is a mold that has been used in Japanese cuisine for centuries and is as safe as any other accepted food microorganism. By leveraging modern technology and safety standards, the application customization and production quality of koji spores are the best they've ever been.

According to Higuchi-san, the best way for koji to be used and appreciated in the future is by experiencing it. Tasting a wide variety of products made from it in dishes, in both traditional and modern cooking, is absolutely necessary. He believes that once folks experience how delicious these dishes are, they'll begin to understand the power of koji. Ultimately, they will gain a greater appreciation of what it can do.

Acceptance of the potential beyond established practices is important. In Japan, koji fungi are still primarily being used in traditional fermented food production, but Higuchi-san would like to see the use broadened further. He feels that the variety of his koji starters will allow makers all over the world to realize wonderful flavors by applying them to local ingredients wherever they may live.

## Koji Preparations

The following koji preparations each have their own intricacies and set of needs to be tended to. We'd like to share our most successful recipes that we hope will give you perspectives on how to apply the old and new starch preparations we have discussed in this chapter. Please consider these instructions as a base whose inoculation and growth guidelines can be applied to practically any other substrate you choose.

# Quick Rice Koji

For the folks making koji at home, there are rare occasions when you're unprepared and need to start a batch of rice koji right away. One really quick way to get the al dente cook is to use a microwave. All you need to do is fill a microwave-safe bowl three-quarters full with rice and add water until the rice is just barely covered. Place a plate on top of the bowl to cover the rice (this will allow steam to escape). Run the microwave at regular time intervals, so the water doesn't boil over. Mix at each time interval. The duration and number of intervals will be based on the condition of your rice and the power of your microwave, but we've found that a total of 12 minutes of cooking at intervals somewhere between 2 and 3 minutes works. If the rice seems dry at later intervals, add a little water. You'll know the rice is done when most of the rice grains don't have a stark white core and the chew is al dente when you test it. You can use the rice right away or leave it covered for 10 more minutes for fuller hydration.

## Popcorn Koji

As we discussed earlier in this chapter, one of our first investigations into the realm of koji making was the idea of creating a starch that was easier to prepare and more accessible than rice. It ended up being something in the pantry that was staring us in the face: popcorn.

The first step is to pop the corn kernels. It doesn't matter how you accomplish it, only that you use as little oil as possible. A hot-air popcorn popper requires no oil, so that's ideal. Make enough popcorn to fill your koji tray to the top. Find a large bowl with plenty of room to mix in. Put the bowl on a scale, tare the weight, and weigh the popcorn in grams. For every 200 grams of popcorn, you will need 20 grams of all-purpose flour mixed with 1 gram of your koji starter. In a small bowl, weigh out the necessary flour and starter, then mix. Start tossing the popcorn in the large bowl while spritzing with a handheld water mister, enough to wilt the popcorn without making it soggy and wet. Sprinkle the flour mixture onto the popcorn and toss until the popcorn is coated. Follow the rest of the standard koji making process, mixing every 12 hours. It should be complete somewhere between 36 to 48 hours later.

## Overcooked Grains Koji (Instant Mirin)

When Jeremy first started growing koji, he was looking for substrates that could be salvaged from other culinary applications, such as overcooked rice used to fortify stock or leftover cooked pasta. Understanding that koji is a fungus and that it just wants to grow and reproduce, he compared it with other wild-fruiting fungi he was familiar with. This comparison led him to realize that the perfectly cared-for rice used in traditional growing wasn't the only viable substrate. Fungi in the wild will grow wherever and however they can. While traditionally prepared rice is the best for allowing koji to reach its full potential, Jeremy realized that when it comes down to it, koji essentially just wants starch and some protein. It will happily grow on substrates that have been overcooked nearly as well as ones that have been carefully prepared.

Jeremy started using overcooked rice that was used to fortify and give body to a vegetable stock. He strained the stock, picked out the mirepoix, and then drained the rice to rid it of excess moisture. From there he inoculated the rice and allowed the koji to grow—and it grew fantastically. One thing that he noticed was that the koji would grow over the surface of this rice so vigorously that a thick matted covering of mycelium would develop, so thick that it could be removed from the top of the rice in one cohesive sheet. Due to the rice being overcooked, the grains mashed together during inoculation and the first stage of growth, preventing the mold's hyphae and mycelium from penetrating too deeply in. The rice under the surface mat of mycelium started to liquefy due to the enzymes the mold was producing and essentially turned into an "instant amazake" that didn't require further processing. If this instant amazake were allowed to sit for a day or two under refrigeration, it would start to weep a syrupy golden liquid. This liquid has a taste, flavor, and aroma nearly identical to the style of mirin (rice wine) called *shin mirin* or *new mirin*. (Shin mirin is virtually free of alcohol and unfortunately has gotten a bad rap. It is viewed by many as the cheapest type of mirin and, due to it containing little to no alcohol, as the one people produce to avoid paying the taxes attached to high-alcohol styles such as *hon mirin* and *shio mirin*.) The "instant mirin" produced from these overcooked grains happens to be nothing like the commercially produced shin mirin, which tends to be highly concocted and more often than not contains additives such as high-fructose corn syrup.

To make this instant amazake and instant mirin, simply use overcooked rice. Inoculate and incubate it as you would with properly cooked rice. After the mold has bloomed, place the koji into a fine-mesh strainer lined with cheesecloth. Allow this instant amazake to drain for a couple of days in the refrigerator. Don't squeeze the instant amazake if you want the instant mirin to be clear and free of residual starch. Use this instant mirin in any recipe that calls for it, such as teriyaki sauce.

## COCOA NIB KOJI

### Sam Jett

*As a follow up to Sam's pH piece, here's the in-depth recipe on how to make Cocoa Nib Koji.*

This recipe is a bit more technical but is worth mentioning here due to its uniqueness. It was a happy discovery utilizing a by-product from a chocolate tart that had a cocoa nib consommé as a base. Not wanting to waste any food, I used the leftover cocoa nibs after making the consommé. Their pH before the process was around 5 to 5.1, but after cooking was closer to neutral. After a little digging I discovered that Charleston, South Carolina, city water was basic; the water that the cocoa nibs absorbed helped to offset the acidity of their nature. This gave me the inspiration to give this recipe a test run.

This make is very similar to traditional koji making, but I am careful not to introduce any more basic additions to the environment. Another aspect is that cocoa nibs don't absorb moisture like rice or barley does, so any excess will water down the process and ruin the batch. Cocoa Nib Koji adds a great level of complexity to miso recipes (my favorite is sorghum-chocolate miso) as well as adding complexity to a shio koji or liquid amino. Think: chocolate shio koji to season a mole!

**1 kg (2 pounds) cocoa nibs**
**3 kg (0.8 gallon) water (Charleston City adjusts their water to a pH of 8.3)**
**0.3 percent koji-kin**

Vacuum-pack the cocoa nibs with the water (you may need to split the batch) and put it in a 140°F (60°C) immersion circulator bath for 1 hour. Strain off the liquid (reserve for other uses or as the liquid component in an amino or miso). Sanitize your hands with alcohol and wear latex gloves for the rest of this process while handling the koji. Sprinkle the cocoa nibs on a tray and let them air-dry for an hour or so. The idea is to get the bulk of the moisture off the exterior, but you will need it to be slightly damp to get the spores to stick.

Transfer the cocoa nibs to the tray you are using and dust with the koji-kin, mixing thoroughly with your hands. Level the koji out and place a slightly damp towel over the batch. Again, be mindful of how this towel was sanitized. Place the tray of cocoa nibs in your rig. Hold at 77°F (25°C) and around 75 percent relative humidity. If you have the setup, place a Wi-Fi thermometer into the batch with an alarm set for 86°F (30°C) so you can be alerted if your koji is getting too hot.

After 18 hours check the batch. Gently stir to provide some oxygen. If the towel has become dry, moisten it again and cover. Check again at 24 hours. Turn the koji again. If you like to make the furrows do so, but don't make the layer too thin. Check the towel and remoisten if necessary. Repeat this process at 30 hours. You should have a solid cake forming at any time after this. Be very mindful of the temperature, as *Aspergillus* produces a lot of heat as it grows. Inoculation should not go past 36 hours. After your cake has grown, cool and dry it rapidly with fans. Break up the cake as needed to lower the temperature.

CHAPTER 6

# Short-Term Enhancement: Quick Koji Applications

Many of the foodstuffs involving koji take fair to considerable amounts of time to make; amino pastes and sauces can take months or even years, as can certain styles of alcohol. Does this mean that in order to enjoy the koji you've just made, you have to wait months at a minimum? No. A wide breadth of applications fall under what we categorize as *short-term enhancement* (STE); they can be implemented almost as soon as your koji has been made and have wide-reaching possibilities.

Shio koji and amazake applications are the most basic short-term enhancements and happen to be the quickest way to leverage koji for flavor. These two methods allow for ease of transforming ingredients just by adding water. This makes the enzyme catalysts readily available to the proteins, fats, and carbohydrates that they're applied to. Shio koji is a slurry of koji, water, and salt, and is typically used as a short-term marinade to create tasty amino acids when applied to foods before cooking, as they're cooking, or even after they're done cooking. On the opposite side of the coin, amazake consists of koji, cooked starches such as rice, oats, or any other grain, and water. This mixture is allowed to incubate or ferment in order to make a sweet and sometimes slightly alcoholic porridge drink. Amazake is a much more versatile ingredient that can be used in sweet or savory applications. This isn't to say that shio koji won't or doesn't work with sweet foods—it does, just not to the extent of amazake.

The methods for using these koji-derived foods for short-term enhancement range from traditional fermentation techniques to updated ones incorporating modern technology such as precision-controlled devices. The traditional applications have been used for hundreds of years in Japan and other parts of Asia. The updated techniques at their core aren't too different from the classical ones. What sets them apart is our current understanding of how to maximize enzymatic activity and leverage it for the best results in any given application. The leveraging of these enzymes is a common discussion throughout this book. At its simplest all you need to do is be able to accurately control the temperatures at which these foods are grown—in the case of the inoculated substrates themselves—or the temperatures at which they are created, as in the making of an amazake high in free fermentable sugars or a shio koji high in tasty amino acids. There are other variables such as salt content, pH, and water activity ($a_w$—see appendix B for more), but the most effective one and easiest to manage in any kitchen, professional or home, is temperature.

## Amazake

Amazake serves as the base for sake and other rice-based alcohols (and by fermented extension, rice vinegar) but is also a star in its own right. Its roots can be traced to early alcohols made from fruits, honey, koji, and rice in regions of what is now known as China. The *Nihon Shoki*, a text written in 720 CE, makes mention of *amanotamuznake*, which has been referred to as an early amazake. There are many reasons why amazake has been a staple in homes for so many years; it's the base for most Asian styles of alcohol and liquor, it is transformative when applied to foods in the kitchen, and it's wholesome and delicious.

When we first started working with koji, amazake was the food that really opened our eyes to the limitless possibilities that koji affords us in the kitchen. It has all the sensually intoxicating elements that fresh koji does, plus it has more versatility and goes further in terms of usability. Over the course of many years, we have used it to do everything from culturing cream, to making butter, to acting as the hydration element in breads and pastries. But the biggest asset that amazake has compared with other koji-derived foods, aside from its enzymes, is that it isn't seasoned with salt. Shio koji, by contrast, is salted and, in our opinion, is therefore slightly less useful. As culinary professionals we want to control the salt in our foods as strictly as possible, and sometimes we don't want salt at all—this is exactly why amazake is our basic go-to in the kitchen. Another benefit that amazake provides is that it is optimized leveraging amylase over protease and for specific applications such as creating sweet treats and confections.

Classically, amazake is made at room temperature and is ready in two to three days. It is a simple combination of cooked starch (traditionally rice or barley), inoculated starch, and water. The cooked starch can be anything from rice to nixtamalized corn, and this holds true for the inoculated starch, too. These ingredients are mixed together and left to be broken down by the amylase enzymes created by the koji, typically at ambient room temperature left uncovered. To accelerate the process, you can set a jar or container in a warm-water bath, easily done with an immersion circulator.

During our early explorations we realized that different styles of amazake could be made to suit different purposes. For example, we made a very watery amazake that we would let sit out and slightly sour for a few days with the aid of our little bacteria friends *Lactobacillus*. This soured amazake proved excellent in applications as diverse as a finishing seasoning drizzled over a plate to a court bouillon alternative when poaching seafood. This sour amazake also works well in dressings, marinades, compression techniques, and a host of other preparations that call for flavorful liquids.

On the other hand, you can make the traditional-style sweet amazake by using half the amount of water used for the sour type. This amazake makes for a great stand-alone treat but can also be used to marinate meat or even be puréed smooth and spun in an ice cream maker to yield a delicious sorbet.

Whether sweet or sour, the uses for amazake are numerous. You are free to use whichever one you want for whatever application you see fit.

For koji applications in general, amylase has optimal activity between 131°F (55°C) and 140°F (60°C) to break down starch into sugar.[1] Although there is a much wider range of factors to optimize, we've found that this works quite well across the board. While simply mixing the ingredients for amazake and letting it sit at room temperature will work just fine, you can accelerate the process by holding it at temperature. This is especially helpful if you want to make a very sweet base for alcohol and vinegar. As you will experience, there are distinct differences between an amazake that is made at ambient temperature and one that's precisely heated.

One thing to keep in mind when using a temperature-controlled device is that these systems aren't perfect. As the chamber/bath/oven is heating, it may overshoot what you've set it to. How much this deviates from your target temperature is specific to the system you're using. To understand how much variation there is, we recommend that you use a supplemental temperature probe to track the temperature. Short of that, setting your system to be on the low side, 131°F (55°C), should work fine. When working to maximize protease enzyme activity for amino acids (umami), the same range of 131°F (55°C) and 140°F (60°C) applies.[2]

Ambient amazake can and will develop sour notes due to contamination from beneficial lactic-acid-producing bacteria. This souring can be used to great advantage, especially if you want to create dishes balanced with sweet and sour flavors. Sour amazake works

## HOW TO MAKE SWEET AMAZAKE

1. Combine 1 part cooked starch, 1 part koji, and 2 parts water.
2. Blend to increase the surface area of the starches.
3. Hold between 131°F (55°C) and 140°F (60°C) for 10 to 14 hours.[3]
4. Store airtight under refrigeration.

fantastically in the creation of breads. It mimics some of the complexity of sourdough breads without the need to maintain a sourdough starter. At Larder most of the breads we produce incorporate amazake into their hydration. When we first developed these breads, we simply replaced a portion of the water called for with sour amazake that had its solids strained out. Sour amazake also serves as a fantastic seasoning for myriad applications—a favorite of ours is to bathe fresh, raw scallops or shrimp in it and serve them as a crudo. It also serves as a well-rounded base for vinaigrettes and even just as a flavorful brightening agent that you can add to a stock or sauce to give it a loving touch. Fruits compressed in a vacuum with either sweet or sour amazake are taken to another level, especially if they happen to be slightly lackluster on their own.

The longer your sour amazake ferments, the sourer it will become. There is also a strong possibility that it will become alcoholic instead of sour. Each of us produces these foods in different environments with drastically different microflora present. We encourage you to experiment with various parameters and variables to find a product that you enjoy. Toward that end, we recommend that once your sweet or sour amazake has a taste and flavor that you enjoy, you store it in airtight containers in the refrigerator. This will slow further fermentation. You can also pasteurize them to halt further development by heating them to 165°F (73°C)—but then you'll lose the enzymatic action each one affords in addition to its inherent taste and flavor. Most of the further culinary preparations that you undertake will benefit greatly from the presence of the active enzymes, so store them cold instead of heat-treating.

## HOW TO MAKE SOUR AMAZAKE

1. Combine 1 part cooked starch, 1 part koji, and 4 parts water.
2. Hold between 131°F (55°C) and 140°F (60°C) for 10 to 14 hours.[4]
3. Strain the amazake.
4. Hold at ambient room temperatures exposed to air, stirring at least once a day, and ferment for up to a week.
5. Store airtight under refrigeration.

## Shio Koji

Shio koji is an incredible food that can provide many of the same benefits as an amino sauce or amino paste but with flavors and aromas closer to those of fresh koji and amazake. As we said at the top of this chapter, salt is the main difference between shio and amazake. Think of shio koji as your go-to flavor enhancer; it's the perfect umami bomb that will keep people begging for more food.

Most shio can be used within as little as a few days, but it really starts to hit its stride after a solid week of fermentation. Again, think of the delicious slice of a perfectly ripened tomato accented with a sprinkle of salt that we mentioned earlier. But take a tomato and add shio koji, with its salt and active enzymes, and the tomato will be so delicious that you won't want to eat one without shio koji ever again. Shio koji will also bring out delicious fermented flavors that can range from the essence of briny sushi rice to the cheesy funk of Parmesan. These complexities in flavor vary depending on the inoculated substrate used to make the shio koji along with the percentage of salt used, the length of fermentation, and the temperature at which the fermentation occurs.

When making shio koji, note that the ratio of water and salt to inoculated grains can be changed to suit your personal preferences. If you want to use less salt, say 2 to 3 percent, then go for it. The same goes for adding more salt. However, keep in mind that fermented foods with salt concentrations under 2 percent can easily become a welcome environment for pathogens. On the higher end, even beneficial, salt-tolerant bacteria and yeasts will be inhibited at concentrations over 7 percent. Should you want to add more or less water to your shio, then do so to suit your needs; some people prefer a shio koji that is a little looser than the above ratio. Always be sure to add your ratio of salt to the combined weight of the water and inoculated grains.

Also keep in mind that the longer you ferment your shio koji, the stronger its flavor will be. There is also the possibility that as its pH lowers and it becomes more acidic, its enzymes can denature and become inactive. As we stated above while discussing sour amazake, the lack of enzymes won't allow you to reach the full potential of a food like shio koji.

Shio koji makes for a great all-purpose seasoning and marinade. From simply adding a splash to some peas when you sauté them to marinating a piece of meat for a couple of days, you'll love what it can do to nearly anything that it touches. It pairs fantastically with protein—partly due to its higher concentration of protease activity from being produced at much lower temperatures than those optimal for making amazake. That being so, we recommend that the grains you inoculate to make your koji have a relatively high protein content, such as barley. We also recommend that you incubate that barley at the lower end of the temperature range, 86°F (30°C), as the koji grows. These two optimizations combined will leave you with a shio koji that nearly rivals young amino pastes in terms of deliciousness.

## HOW TO MAKE SHIO KOJI

1. Combine 1 part koji and 1 part water.
2. Weigh this mixture and salt to 5 percent of the total weight.
3. Blend to increase the surface area of the inoculated starch.
4. Transfer the ingredients to a non-reactive container, such as a glass mason jar.
5. Hold at ambient temperatures exposed to air. Mix the shio koji once or twice a day for up to 7 days.
6. After 7 days your shio koji is ready. Store it in an airtight container in your fridge.

## PLAYING WITH SURFACE AREA

When it comes to short-term enhancement, the amount of contact you have with the liquid koji is key to yielding amino acids. As a result, if you increase the exposed surface area, you can accelerate the process to make a protein more flavorful. We feel that this is best explained through examples. Who better to present this information to you than chefs who have come up with ideas beyond what we could ever imagine?

### HOW THE SHIO KOJI BURGER WAS BORN

**Alex Talbot and Aki Kamozawa**

*Two of the most inspiring food tinkerers we know are Alex Talbot and Aki Kamozawa, who started their food blog* Ideas in Food *in 2004 and followed it up with the critically acclaimed book of the same name. Their posts offer insight into the wonders of cooking beyond the standard techniques, and they grew into books that really made you understand the whys. When they began sharing, they were among the few food writers who were demystifying the intricacies of the relatively newly harnessed science and food technology techniques being used in fine dining restaurants. They were doing it at a time before foodies knew what hydrocolloids and sous vide were all about. They showed us that critical thinking was important to truly understanding how to cook and bake. We were honored when they invited us into their kitchen to share knowledge and experiment with koji.*

Koji is a subject that has long fascinated us. One of the issues we have with shio koji is that while it adds umami and complex flavors to a dish, a universal flavor profile comes with it as well. When you add koji to various dishes, they all have a thread of familiarity to them because most koji is fermented from white rice. During a workshop with Rich Shih, we wanted to see what happened if we used different grains to grow the koji. To this end we used three kojis that had been fermented from jasmine rice, brown rice, and barley to make our shio koji with an equal weight of water and 10 percent salt added to the mix.

We decided that burgers would be the perfect vehicle for testing the seasoning. We began with one batch of ground beef, and seasoned one burger with a 10 percent salt brine. We made three more burgers and seasoned each with a 10 percent addition of the three different kojis. (We decided upon 10 percent because each liquid seasoning was 10 percent salt and therefore each burger would end up with a 1 percent salt content and be equally seasoned before cooking.)

After mixing up each burger, we put them on a rack in the refrigerator overnight. We wanted to give the enzymes time to work their magic on the meat. The burgers were uncovered and developed a dry layer on the outside of each patty. This proved to be a bonus because it seemed to keep the meat from sticking to the grill.

The next day we cooked the burgers. The first immediate result was that the burgers seasoned with koji browned more rapidly than the brined burger. We used the finger test to check for doneness, which proved to be an error because the outside layer of the meat was firmer from the air-drying, leading us to believe the burgers were medium rare when in fact the interior was still quite rare. But we didn't let this deter us from tasting.

We were struck at once by the range of flavors in the meat. The brined burger was seasoned, meaty, and juicy. The jasmine koji had added a sweetness to the meat. The brown rice koji burger had surprising notes of blue cheese. The barley koji added an intense flavor of porcini mushrooms. We were struck by the clear differences in flavor and the discovery that we could create new flavors, based on the koji grain, that could be tailored to what we wanted to cook. It opened up a whole new world to explore.

Over time we've played with several other grains, including emmer and Tamaki Gold rice. We've also cold-smoked our homemade shio kojis to add another dimension of flavor. For us the goal is always to create a koji that enhances the ingredient it's paired with, without overpowering it. All kojis have very distinct flavors, and it's easy for them to take over the spotlight. Our goal is always to weave them into a dish so that the koji enhances the overall flavor.

## RECONSIDERING KATSUOBUSHI

### Jon Adler

*Jon Adler ran Sevenstrong restaurant in Northampton, Massachusetts, with his sister Katherine, and is now a sommelier at Single-Thread. The core concept of Sevenstrong was to be serious about sourcing locally. One key element of that was whole-animal butchery and using every last bit.*

We always found creative ways to cook the off-cuts like beef skirt steak, flank steak, and the secreto or spider cut. However, in a small restaurant with limited staff, you don't always have the time to dedicate to a special. We decided to take these pieces of meat and age them so there was significantly less moisture content; this would allow us to shave them very thin against the grain on a meat slicer, leaving a product that resembled katsuobushi. We took shio koji and sprayed the meat shavings with it, leaving them in the open air overnight. This would allow them to dry even further, becoming seasoned by the salt in the shio koji; most important, the enzymes from the koji would free amino acid chains in the beef proteins, giving us an amazing umami delivery system. Here is our recipe.

**Flat untrimmed off-cut of beef, pork, or whatever tough cut you like**
**Strained shio koji in a spray bottle**

Set your meat on a rack in a refrigerator and place in front of a fan. Leave it for 2 to 4 weeks or until it has the dryness of leather or jerky. The water activity does need to be below 0.85 in order to ensure there is not enough moisture for harmful bacteria to reproduce. This is on the order of 30 percent weight loss. Slice paper-thin on a meat slicer. Place on a parchment-lined tray and spray with shio koji until fully coated. Leave overnight, then refrigerate.

CHAPTER 7

# Amino Pastes

Amino pastes are what we call various high-protein-based condiments that are autolysed and fermented, *amino* due to the high level of tasty amino acids that these preserves contain, *paste* because that's the best physical descriptor for these foods. You may already be familiar with two such traditional savory ferments: Korean gochujang and Japanese miso. Both have immense cultural significance and have been codified, produced, and consumed for centuries. Their continued growth in popularity across cultures is driven by their deliciousness. That's exactly why we are so interested in inspiring you to create your own.

Amino pastes are used to add a depth of flavor that no other condiment can. Their traditional uses cover a wide range of applications: They serve as soup and sauce bases, marinades and pickling ingredients, as well as seasoning and fortifying agents. In terms of their versatility in cooking, think of them as the equivalents of butter or eggs. Each style of amino paste has its own specific use but can also be used in a broader manner and easily be swapped for another. Bottom line, they're all packed with umami and can make just about anything taste better when applied in appropriate amounts. Here are some more reasons that amino pastes have been so important: They're more delicious than any other plant-based condiment; they're very inexpensive compared with other sources of protein; they're shelf-stable and don't require refrigeration; they're incredibly dense in macro- and micronutrients; and they're easy to make.

Different styles of amino pastes are reflective of the terroir, available ingredients, and peoples from whose lands they originated. China is the original progenitor of all amino pastes, which is reflected in its huge number of diverse starters and final products; the most basic culture alone, known as *qu*, can be made in more than 120 different variations. Korean amino pastes are mainly made using a mixed culture of various bacteria and mold, specifically species of *Aspergillus*, which are then dried and fermented a second time in brine. Japanese amino pastes are made primarily with koji and then allowed to ferment. What all of these have in common is that they start with high-protein ingredients, salt, and a culture, and then are allowed to age for various amounts of time.

These more generalized styles can be transformed into wildly amazing and captivating variations. As we stated previously, any ingredient with protein can fuel the umami. Traditionally the primary protein driver in amino paste is soybeans. During the course of our experimentation, however, we have used other legumes, nuts, seeds, and everything from whey powder, cookie dough, bacon, and cacao beans as flavor drivers. One of our standout discoveries is a ricotta-based miso that developed aged Parmesan/Romano cheese flavors in two months—something that typically takes a year. In short, anything with protein will work as an amino paste base. The possibilities are only limited by your imagination.

As a result of years of experience, we've established simple and straightforward steps to yield a great product. Don't be intimidated by these specific guidelines; as you make an amino paste, you'll fully understand their importance. We primarily create amino pastes in the generalized style of miso because it's a simple process that yields wonderful results (and of course koji is the magic ingredient for developing depth of flavor, or umami).

There are only three ingredients required to make miso: koji, a protein base, and salt. In traditional miso making, the ratio of koji to base to salt is driven by flavor and fermentation duration. For short-term (two weeks to three months), "light" miso leveraging, sugar-fueled fermentation, you need a higher koji and lower salt percentage; for long-term (six months to several years), "dark" miso driven by umami, you need a higher-protein base for amino acids and more salt to protect the ferment from going bad. In our experience, a simple guideline from Sandor Katz's *The Art of Fermentation* works well. Short-term requires 1 part koji to 1 part base by weight, plus 6 percent salt against the total (we typically use 5 percent, however). Long-term requires 1 part koji to 2 parts base by weight, plus 13 percent salt against the total. The koji and protein base choices are entirely up to you.

If you let a short-term amino paste go too long, there's less protein for umami and a small level of risk that your amino paste will not be good after the extra time; however, we've never had a problem with this. The key to success is proper containment, which we will be discussing later in this chapter (see "Mason Jar Containment" on page 125 and "Custom Fermentation Lids: A Practical Salt Solution" on page 126). The light amino pastes that we've let go for at least a year are complex in their own way and are much easier to use due to their significantly lower salt content. All in all, we suggest you start with the recommended ratios for the long-term amino pastes, and then consider other possibilities once you are comfortable and confident with the projected outcomes.

This might seem obvious, but it's worth reiterating: Amino pastes are *pastes*. This means the base ingredients are mashed or processed into tiny bits to optimize conditions for fermentation and inhibit bad microbes that would otherwise make you sick. We highly recommend that you process at least half of your mixture into a purée for good results. If you don't, there will be small pockets of air trapped within the mash, which creates not only the potential for spoilage but also an environment for pathogenic microbes to thrive. The puréed portion fills in these potentially hazardous voids. Aside from these health risks, the finer the paste, the easier it is for enzymes and favorable microbes to access their respective fuels to create a tasty product. If you want your amino paste to be completely smooth, then purée it both once it's finished and when you intend to start using it.

Water content is another important factor for proper fermentation of amino pastes. Sometimes the mixture can end up dry if you're using dried koji or trying something new. Also, you may want to make a looser paste so you can take advantage of the liquid that separates out. (In miso making, that liquid is called *tamari*, which is a close cousin to soy sauce and can be used in the same ways.) If the final mix feels wet to the touch, that's a safe guideline for the appropriate amount of moisture content. If it's not quite there, feel free to add water until it's right. Water is the baseline, but the liquid can be anything else that makes sense to include. Just make sure to weigh whatever is added so you can maintain the salt percentage of the make you've decided on.

Preparing the mix for the months to years of fermentation is a straightforward and very important step. Do your best to pack your amino paste into a non-reactive vessel, such as one made from wood or glass, as tightly as possible. One traditional way to do this is by hand-forming and compressing balls of the mixture and throwing them into the fermentation vat.[1] The size of the ball is what you can comfortably compress between your hands, very much like making snowballs. You can tell you've used too little water if the ball cracks when flattened. The spheres are then thrown into the vessel to compress the mash further and layer until the container is practically full. As much fun and effective as this process is, though, it's not entirely necessary. We've found that you can still get good results by just filling and compressing in layers without going through the trouble.

Once the vessel is mostly full, you want to create a seal that compresses and isolates the ferment from the outside environment. This is accomplished by having a non-reactive cover, which can be as simple as a plastic bag filled with salted water or salt or weights made from glass or stone, that just fits inside the vessel and comes directly in contact with the material. The cover needs to be held in place, most commonly by weight, so it prevents the carbon dioxide pressure of the fermentation process from creating air gaps within the mash. But there should be a small gap between the inside of the container and outside of the cover to allow this pressure to escape. For additional protection, sprinkle a layer of salt on the top of the mixture before you put the cover into place. This creates a high concentration of salt to ward off any unwanted microbes for a short period, ideally until the fermentation catches up and makes it uninhabitable for unwanted microbes. A layer of plastic wrap against the surface can be used in conjunction with the cover to ensure a good seal against the top.

## Mason Jar Containment

For those who are just starting to make amino pastes and want to use something readily available, we have developed a method for accomplishing compression containment with standard mason jars.

**MATERIALS**

Miso ready to be contained
Widemouthed mason jar (16, 32, or 64 ounce)
Plastic or metal regular-mouthed lid outside the top of the ring inside a sandwich bag
Salt
Plastic wrap
Metal widemouthed lid and ring

Compact the miso into the mason jar as you normally would to get as much air out as possible. Fill until the compressed paste line is at the level seen in the image to the right as the thin salt layer. Check the level by placing the regular-mouthed lid/ring upside down on top of the miso. The ring must stick up just past the opening of the glass jar to create the surface pressure needed for containment. Add or remove miso to get to the appropriate level.

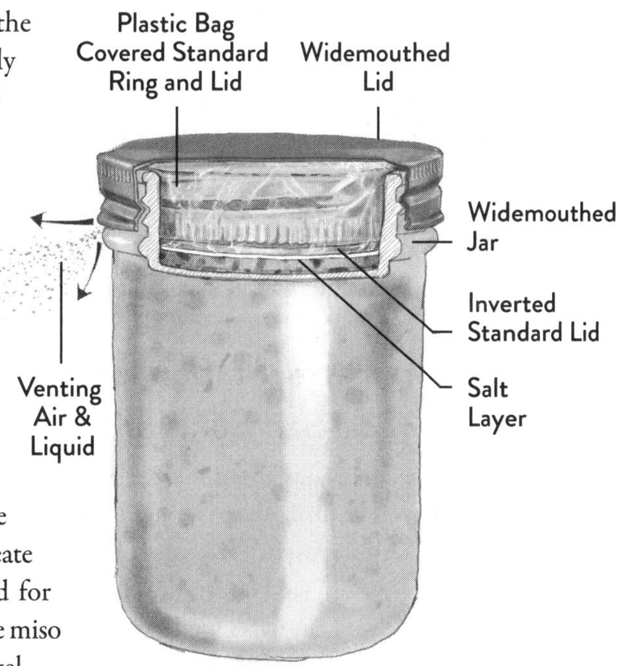


Mason jar amino paste containment. *Illustration by Max Hull.*

Once you reach the correct level, remove the lid/ring and

slowly pour in enough salt to cover the exposed miso. Tilt the jar in multiple directions until the entire surface is lightly covered. Dump out the remaining salt. Put two layers of plastic wrap over the mouth of the jar with at least 1 inch (2.5 cm) of excess all around. Compress the regular-sized lid upside down into the plastic wrap at the mouth of the jar until it is in contact with the miso.

Gather all of the plastic and bring it into the center of the lid, then put the widemouthed lid over it. Push the lid down onto the jar as you tighten the ring until it just engages and compresses the miso. If you're watching how much, it's about an eighth of a ring rotation. If you can't engage the ring, you will probably need to remove a small amount of the paste.

When you're done, the top must *not* be sealed. There has to be a gap for the future carbon dioxide created to escape or the vessel will pressurize and potentially explode. When this is done correctly, you do not need to worry about any undesirable microbes growing on your miso. We have never had a problem when using this technique. The one minor drawback to this technique is that the tamari (liquid) will start coming out of the jar as it ferments, so you will have to account for that. We typically put all of our jars on a large non-reactive tray or bin such as CorningWare or a plastic tub; then, when the liquid escapes, it makes a contained mess that's easily cleaned. We keep the jars wherever we can keep an occasional eye on them, which sometimes is on the kitchen counter or in a clean area of our basements; essentially, you can keep them wherever you are most comfortable.

## Custom Fermentation Lids: A Practical Salt Solution

Once you get serious about making amino pastes, you'll find that there are limited options for finding the right vessel to handle the compression containment. Also, most of us like the versatility of using whatever non-reactive vessel we have on hand or can find.

Miso barrels and vats are designed with a lid that is just smaller than the opening of the vessel while in contact with an amino paste. The lid is sturdy enough to support the weights or stones above to maintain the compression during fermentation. This maximizes contact area and maintains

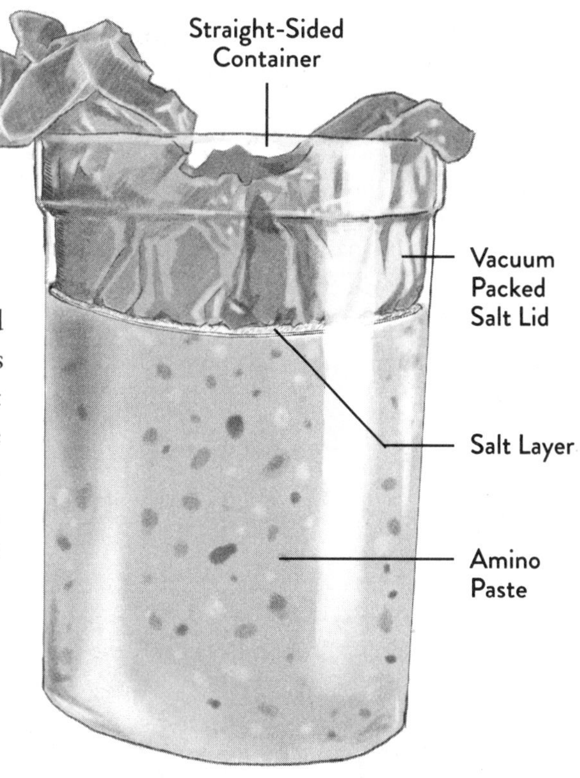


Custom vacuum-packed salt lid for amino paste containment. *Illustration by Max Hull.*

the gap that allows carbon dioxide to escape and liquid to overflow above the lid.

Some common solutions for covering and weighting the top of an amino paste (aside from a fermentation barrel and matching lid) include weights for mason jars and crocks that don't quite cover the entire surface and require supplementing, and, more commonly, plastic wrap. There's always the option of using a bag of salted water, which when lowered into the top of a vessel can match the vessel's shape. However, bags of water are fairly limited by their density, tend to be difficult to keep in place, and aren't friendly when it comes to adding weight. These solutions are just work-arounds for a custom lid, and they're not great.

There is a way to practically (and inexpensively) form a custom lid. The lids used in traditional miso fermentation barrels are cut to size. What if you could use just three things to accomplish the same purpose on practically any vessel with much less effort? It's possible! All you need is a vacuum sealer, a vac bag roll width that is at least a couple of inches larger than your vessel opening, and table salt.

As we all know, a box of salt is fairly heavy for its size. The density of salt is 1.24 oz/in$^3$ (2.15 g/cm$^3$), which is more than twice the density of water at 0.59 oz/in$^3$ (1.02 g/cm$^3$). (We chose table salt because it's fine and can be compressed to eliminate as much air as possible.) A quick water displacement test to determine the volume of a vacuum-packed salt lid weighing 3.25 pounds (1.47 kg) puts us pretty much in the middle at 0.87 oz/in$^3$ (1.51 g/cm$^3$). We found out quickly in our initial experiment that kosher salt does not compact as well; you pretty much end up with the same density as water, so it floats. The reason why we want the cover to be denser than water is so that it will sink if you're using it to hold the solids in an amino paste below the brine that accumulates on the top.

To make your lid, you first need to make the mold for the salt to fill. Measure 3 inches (7.5 cm) down from the top of the fermentation vessel and mark in a few locations with kitchen or masking tape. This assumes that

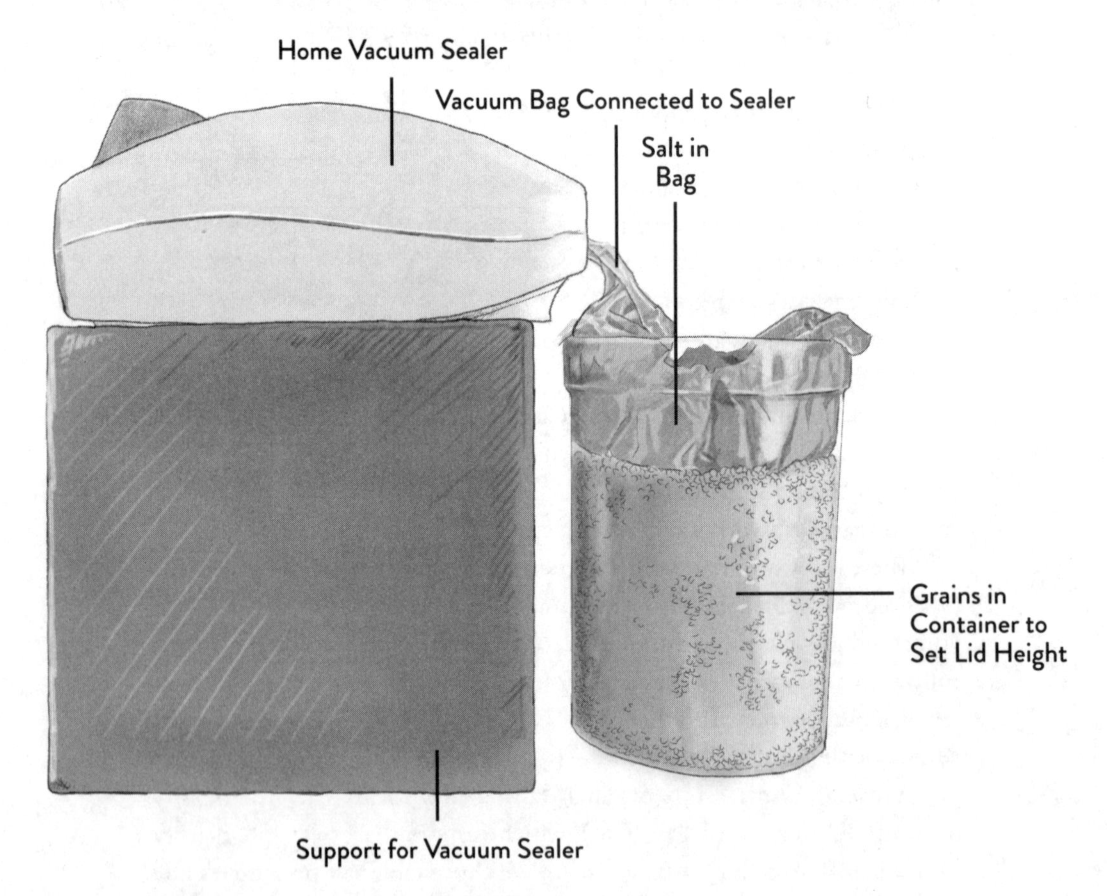


Setup for making a salt lid. *Illustration by Max Hull.*

you'll have 2 inches (5 cm) of sealed salt and 1 inch (2.5 cm) of headspace. Fill the vessel with any grains to the tape mark. Next, determine the length and width of the vacuum bag roll you'll need to cut. Position your vacuum sealer so the opening is just above the vessel. The closer you get, the less bagging you'll need. Pull the bagging out until you are confident that you can fill salt on top of the grains to a thickness of 2 inches (5 cm) and be able to vacuum-seal the bag. For vessels that have slightly smaller openings or are quite large, you'll have to make a two-piece lid split at the middle. (We're sure you can do it in more pieces, but we haven't had to.) Cut and seal one side to make the bag. You know what to do from here. In some cases you may need to use the handle of a wooden spoon to get the salt to fill in the edges.

As a side note, if you want to use this to weigh down ferments like kraut, kimchi, pickles, or other applications, you first need to get your solids below the brine. This means you just have to increase the gap between the cover and vessel opening. You can do this by cutting a long 2-inch-wide (5 cm) strip of cardboard or whatever is flexible and gives you the gap you want and cutting its length so it can be placed as a collar inside the vessel above the grains. This strip of material will make the salt lid smaller by the thickness.

With regard to larger weights, the only ones we know of at this point are the split doughnut-shaped weights made for lacto-fermentation crocks. There are ceramic and glass versions, as well. For a direct comparison, the custom salt weight we made in the circular vessel here is the size for which you'd use an off-the-shelf 1-gallon (3.8 L) split doughnut weight. Let's say the ceramic weight costs $20 and weighs 2.5 pounds (1.1 kg). That puts us at $8 per pound ($18/kg). Assuming you have access to a home vacuum sealer, which is reasonable these days, here's the breakdown. The cost of a standard 26-ounce container of table salt is just under $1. Two containers coincidentally put us at the size we needed, so that's $2. On top of that, let's say the bag costs $0.50. The salt lid is $2.50 of materials and weighs 3.25 pounds. That puts the salt solution at $0.77 per pound ($1.70/kg): ten times less than the doughnut weight. Also, as we mentioned before, split doughnut weights are not ideal for keeping all the solids below the brine and thus aren't the ideal answer for containing a miso.

If you're new to fermentation, you're probably wondering why we're focusing so much on containment. Simply put, it's important to the quality of your final product. Also, there isn't one solid solution that covers all fermentation vessels. We've been annoyed by the lack of good options and we think this is a pretty great solution. Yes, there are issues that come along with using vacuum-bagged salt. The big one is that if the plastic is punctured, it's game over. But keep in mind that puncturing the bag isn't easy to do; vacuum bags are made to take a beating. Think about all the rough handling endured by vacuum-packed products that are frozen, piled up, and transported every day. Of course, there are things you can do to ensure it won't break. Bagging a second time is the first thing that comes to mind. Put on a plate or tougher insulator to take the beating if you want to add weight on top.

Note: If you have access to a commercial vacuum sealer and want to do this with standard square restaurant-style containers, use the shorter containers with an opening the same size as the mold to make the salt lid.

## Monitoring the Fermentation Process

Depending upon your containment setup, small air pockets may form as your amino paste ferments, but these are of little to no concern. These air pockets are carbon dioxide, produced by favorable microbes. (While an air pocket devoid of free oxygen may be of concern due to botulism, the pH and salinity of the amino paste at this stage are inhospitable to the production of the botulinum toxin.) Another consideration during the active fermentation stage is that the liquid may push out of the vessel if the cover is close to the top. As previously mentioned, this is easily solved by placing the vessels on a clean tray to prevent a mess.

So how long should your amino paste ferment? Honestly, until you like how it tastes. In general, you can never age an amino paste for too long. There's potentially the issue of rancidity for higher-fat applications, such as amino pastes made from meat, should they age for too long, but this is less a concern for safety and more a concern of quality, as rancid fats taste disgusting (see "Is Eating 'Rancid' or Oxidized Fats Bad for You?" on page 138). Some styles of amino pastes are allowed to age for years before they're consumed and enjoyed, and generally speaking the older a paste is the more we enjoy it due to the complexity and intensity that develops over time. You will see a miso's color darken as a result of a long-term Maillard reaction, the same thing that occurs when searing a piece of meat.

As a general guideline, we recommend that you age amino pastes for at least six months. The key reason for this is to allow proteins to be broken down into amino acids for umami. Anything shorter and the miso is dominated by fermented flavors, which isn't necessarily a bad thing, but it doesn't allow for enough time to develop the depth you are typically seeking in this application. At the end of the day, let your taste buds decide.

With all this said, there are a couple of exceptions to aging an amino paste for six months. The first is for amino pastes made from cooked meats. Whether you are using a sausage, bacon, or burnt ends from a brisket, you should realize that the various cooking methods used to make these foods have already worked to denature their proteins. Once koji is added the whole process is essentially sped up considerably, so much so that just by mixing in the koji you already end up with a product that is a serious umami bomb. These meat-based amino pastes tend to obtain the deep full-bodied flavor that aged styles have in a fraction of the time—two to four weeks as opposed to six months or more. The other major exception is amino pastes made with *A. sojae*–cultured beans. This style uses beans that are cultured instead of a koji-cultured grain mixed with cooked beans. This bean amino paste is fantastic and ready to use in as little as four weeks.

## Adding Complexity Ahead of Time

In terms of non-traditional ingredients, you can use techniques you already know to condition the base to be better suited for the process. Think about how the cracked wheat in a traditional soy sauce base is toasted, for instance; it's not a reach to use heat to add great depth.

Sarah Conezio and Isaiah Billington at White Rose Miso treat a few products for their delicious non-traditional miso makes. They like to manipulate the base ingredients, for the sake of flavor and to control the amount of moisture, before combining with koji for an amino paste. They do this by toasting ingredients such as nuts and seeds, or roasting ingredients such as sunchokes or sweet potatoes. Doing this creates great depth of flavor that is carried through the aging process.

## What's Happening as the Paste Conditions

As an amino paste ages, it *lyses*, meaning enzymes break down the fats, carbohydrates, and proteins into the various simpler compounds that are the building blocks to make it delicious. Fermentation differs from autolyzation in that it is microbially driven rather than enzymatically driven. But the autolytic and fermentation processes work together well when they're warm. Therefore, we like to keep our amino pastes between 60 and 100°F (15–37°C) to maintain a solid level of activity. Any lower and the paste becomes dormant; any higher and beneficial microbes are killed off. You shouldn't have any trouble finding a place in your kitchen that falls within this temperature range to age your paste. You can hold an amino paste at 140°F (60°C) and solely rely on the enzymes present, but we feel you lose the depth that the fermentation microbes bring. They cannot survive this high a temperature. Holding foods at such a temperature also uses a lot of resources, including electricity, and isn't necessary for most of us.

Funky things can happen to your amino paste as it's fermenting. It may attract flies. Mold might grow on exposed surfaces. Salt crystals may nucleate and precipitate out of suspension. These are all normal and shouldn't be cause for panic. If you have contained your amino paste properly, these conditions will be isolated to what you can see and are easily removed, leaving the remaining paste perfectly usable. When it comes to potential spoilage, troubleshoot it like any other food with your eyes and nose. If it looks and smells bad, it probably is. Of course, you may be unfamiliar with conditions that are acceptable, especially when you first start making. So don't be afraid to ask someone who has more experience for guidance. This is exactly why we use #KojiBuildsCommunity on social media forums—search for it and we're confident you'll find examples from many other individuals addressing your questions. A lot of time and energy have gone into your amino paste, so try to check in with someone before you just toss it and risk wasting something perfectly delicious.

Once your amino paste is where you want it, you should either refrigerate it or store it properly contained in a cool dark place. If you really enjoy the taste and flavor of your amino paste, you may want to also consider cooking it to stop the microbes in it from further transforming it. All you really need to do is raise the core temperature to above 165°F (73°C) to halt nearly all activity. This can be done on a stovetop using a pot over low heat and constantly stirring, on a baking sheet in a low oven, or by bagging the paste and cooking it sous-vide (sealed in an airtight container and cooked in a temperature-controlled water bath). Generally speaking we prefer using sous-vide for this over the other methods, as it doesn't require constant supervision and you don't dry out your amino paste. Circulating a vacuum-bagged amino paste at 190°F (88°C) for two hours is all you really need.

You'll notice that your amino paste will take on a darker color and much deeper taste and flavor after cooking. The paste will now also be incredibly stable and cease to further transform in taste, flavor, texture, and color. The downside to this is that if the applied uses you have in mind for the amino paste require you to have live microbes and active enzymes present, then you'll be out of luck. In this case keep the amino paste cold when storing to keep it in stasis instead of heating it.

## An Ultimate Application Mind-Set

When it comes to applications, amino pastes have way more potential than has traditionally been explored. We've put them into chicken noodle soup, burgers, tomato sauce, pierogi, hummus, cookies, pies, jam, and chocolate ice cream, to name just some of the successes. An all-time favorite of ours is a compound butter made with amino paste, which we use for everything from sautéing vegetables to schmearing on a bialy. The beauty is that the paste is so concentrated and has so little moisture that you don't need to adjust a recipe in most cases. There are reasons that miso has become ubiquitous around the world: It's versatile, relatively inexpensive, and scrumptious. The key is to think of amino pastes as salt with a greater depth of flavor.

In this section we offer some amino paste recipes to get you started. Because we want you to be free to make whatever you want and not be bound by a recipe, consider these few adventurous makes as starting points to spur your future creativity.

# Basic Amino Paste

This is the master recipe that we use for nearly all our amino pastes, whether we make them from soybeans to beef heart. We've kept it simple and straightforward so that you can easily memorize it. It can be scaled up or down as you desire by either multiplying or dividing the ingredients. You can make 20 pounds (9 kg) just as easily as you can make 2 pounds (900 g). Try to take into account how much of the paste you plan on using and adjust the measurements appropriately. For both light and dark basic amino paste, fresh koji is preferred, but if you're using dry koji, mix in 25 grams (about 1 fluid ounce) of lukewarm water into the koji in a small bowl and allow it to hydrate for a couple of hours at room temperature. If you don't want to wait, process into a rough paste.

*Light (2 weeks–3 months)*
**250 g koji**
**25 g kosher salt**
**250 g protein**

*Dark (6 months–1 year+)*
**165 g koji**
**65 g kosher salt**
**330 g protein**

*Note*: For ferments applications, we recommend using grams to measure rather than their US equivalent. This assures an accurate salt percentage for food safety.

Add the koji and salt to a medium mixing bowl. With clean hands, mix the koji and salt together so the latter is evenly distributed. Now combine mixing and squeezing the koji and salt together to break down the koji into a paste as much as possible. Don't worry too much about making it super fine or missing some grains; the pieces will have the opportunity to break down fully during the fermentation process.

If the protein is in a solid form and cannot be simply mixed into a paste, process accordingly. Most ingredients can be cut up into chunks and run through a food processor.

Add the protein base and mix thoroughly. Pour the contents into any non-reactive pint container. A mason jar is preferred. Store the jar at ambient temperature for the specified time for either light or dark miso.

# Roasted Whole-Squash Miso

We wanted to make a miso powered by single vegetable that had enough core flavor to warrant waiting several months to a year. We decided to use everything from the small 898 squash from Row 7, the first seed company built on chef-and-breeder collaboration, led by chef Dan Barber of Blue Hill at Stone Barns and Michael Mazourek of Cornell University. This particular squash is special because it was bred to be delicious rather than to excel at the volume production most breeders are concerned about. You may know it by the first iteration, the Honeynut, a refined butternut squash built on many generations of breeding work.

When you think about the sweet celebrated flesh, you might be asking why we wouldn't make a nice amazake or kojizuke with it instead. The answer is pepitas, aka squash seeds, which are a mainstay in Mexican cuisine. When we first started our koji journey, we made a pepita amino paste with teff koji that we loved. It only made sense to incorporate the seeds into the squash flesh for an amino paste. As we've said before, you don't need to be smacked in the face with a ton of umami; it's the subtle complexity that elevates the flavor in such a way you don't even really know it's there. All you taste is the key component.

Whenever squash is in season, restaurants drown in squash innards. The easiest way to utilize them is to roast, season, and eat. However, with the fibrous hulls intact, most of us don't find them to be that desirable. Cracking large quantities to supply a restaurant for service is not practical. Did you know that you can roast and grind squash seeds, hulls and all, to make a tahini-like paste? Did you also know that you don't need to separate the strands of squash attached to the seeds? In fact, it adds great flavor when you roast them with the seeds. When it comes to full utilization of a product, let's not stop there. Squash skins are normally removed and thrown out due to their toughness. But we've found that when you slice squash with the skin intact into wedges and brush the skins with oil, you get a browning that creates a depth of flavor when added to a miso. No waste also means less work!

# Whey Powder Amino Paste

One thing that we've always wondered is how much protein can you load a miso up with to create umami. In all our history of making umami-rich condiments, we don't believe we've ever maxed out the protease enzymes in a mix. Maybe we just didn't have enough protein? To challenge this assertion, we decided to seek out one of the most concentrated protein media—whey powder. It has little carbohydrate or fat, which means it's a wonderful fit for an amino paste. Outside of protein jacking, we've worked with a number of dairy products in the past with wonderful success.

To make this amino paste, all you need to do is add enough whey powder to a batch of koji so the grains are coated when mixed. After that, add just enough water to create a paste the consistency of miso. Add 5 percent salt by weight and mix. Contain as previously laid out and allow it to age for 2 months at ambient temperature. This will yield an intense nacho cheese–flavored paste.

## HIGH-FAT AMINO PASTES

Before we move further into this concept, we'd like to address any concern about using high-fat ingredients in an amino paste. The fear of consuming rancid fats has led to an aversion to making amino pastes on a high-fat substrate. Rancidification of fat can occur through oxidation, irradiation, enzymatic lipolysis, and heat, with some of these reactions being initiated by certain microbes.[2] Rancid fat often emits a detectable off-smell, but this is not always the case. There are numerous animal studies showing that consuming rancid fat leads to serious health risks, leading to deficiencies in vitamins or toxicity.[3] Very few studies have been conducted on rancid fat consumption in humans, with only several severe case studies.[4] However, rancid fats can be found in the human diet, and some cuisines include dishes where the fat has *purposely* been made rancid. Given the right conditions, of course, all fats will become rancid. Therefore, during fermentation, it's best to try to slow down the process by which fats become rancid, such as by maintaining your ferments in dark containers with little exposure to oxygen and then storing them in a cool place once fermentation has completed.

We have experimented with avocado and other high-fat amino pastes like cheese, and they're delicious. We do a 1:1 ratio by weight of the high-fat ingredient to rice koji and add 5 percent salt to the total. We've held them for as long as a year and then contained them like miso with very limited exposure to oxygen. But although we haven't experienced rancidification making such products taste bad, there's a definite risk. To be on the safe side, we recommend fermenting in the refrigerator.

We've also made fantastic high-fat amino pastes from leftover mashed potatoes, and others from bacon. In the case of the mashed potato amino paste, we added vital wheat gluten or whey protein powder to boost the protein content. If you don't add these, you run the risk of making an unpleasant amino paste that tastes of curdled milk (from all the butter and cream) and rubbing alcohol. In the bacon experiment we exercised the controls mentioned previously and have never had a problem. In fact, the bacon amino paste made at Larder sells so fast that we often have trouble keeping up with its production.

## IS EATING "RANCID" OR OXIDIZED FATS BAD FOR YOU?

**Johnny Drain**

*Johnny Drain is the head of the fermentation research and development at Cub, a forward-thinking restaurant in London. He is also the co-editor of* MOLD, *an innovative magazine about designing the future of food. Johnny is well known for digging deep into the science of food and fermentation. Here's an expansion on an excerpt from a Nordic Food Lab piece he wrote that will help assure you that fermenting with fats can be done safely.*

Let's start by defining rancidification as the breakdown of long-chain fatty acids into shorter-chain, stronger smelling and tasting fatty acids. This is conceptually much the same as carbohydrates breaking down into simple sugars, or proteins breaking down into amino acids. It's just another way of making flavor. Fatty acids themselves are neither unpleasant nor dangerous. Short-chain fatty acids are found commonly in many foods that we eat, but at high concentration they can become pungent and challenging; for example butyric acid is found in levels that contribute pleasingly to the flavor profiles of hard cheeses such as Parmesan, but at high concentrations, such as in very old butter, can taste like baby vomit! All in all, it's about understanding the processes at play so you can reach a delicious peak flavor, which takes a little more attention with fats than with protein and starch breakdown.

There have been a number of alarming claims about the harmfulness of rancid fats, linking them to, among other things, cancers. However, much of the research we found cited alongside such claims has been extrapolated or is only tangentially relevant—for example, experiments performed on animals in which, in order to produce measurable levels of adverse effects, much larger amounts (in relation to body weight and life span) than would be encountered by humans were used. Tom Coultate highlights this fact in *Food: The Chemistry of Its Components* and adds that "the possibility that lipid oxidation products are toxic to humans remains unresolved" and "[despite claims to the contrary] there is little evidence at the present time to suggest that oxidized fats do cause cancer in humans."[5]

Furthermore, McGee reassures us in *On Food and Cooking* that "rancid fat won't necessarily make us sick, but it's unpleasant," and, as mentioned, the processes through which we make some cheeses, and the compounds these processes produce, are analogous to those that cause rancidity in fats.[6]

Therefore, in essence, the answer to the question is: no, not in the amounts one might typically consume them as part of a balanced diet.

## Sunflower Seed "Douchi"

Our method for sunflower seed "douchi" was inspired by the process of making traditional douchi, or Chinese fermented black beans. The interesting thing about how douchi is made is that the extended period of soybean incubation ensures plenty of mold infiltration. As a result, there is a lot of spore generation, which can create an undesirable flavor. To prevent this, the beans are gently rinsed to remove the spores.[7] This made us think about the constant battles with sporulation when we first started making whole-grain koji with the bran intact. Even with pearled barley, there's always the risk of stressing the koji into a state of reproduction. Why not treat it in the same fashion as a douchi? You have the enzymes you need, so why not rinse the grains as you would the soybeans?

The theory seems sound, but the hitch is that grains are primarily starches with a small percentage of protein. Since we're driven by umami here, we'd need to fix that. We started to think about how we could make better use of the protease enzymes. Why not follow a soy sauce method? Even though we didn't have the traditional soybean and roasted cracked wheat koji mix, we figured that brining a protein with koji would work. If you think about it, shio koji is kind of sort of a short-term amino paste or sauce.

We decided to experiment with roasted sunflower seeds. We mixed equal parts by weight sunflower seeds and rinsed sporulated barley koji then added 5 percent salt against the total. This mixture was placed in glass jars to which a 5 percent salt brine was added until it covered the mix by 1 inch (2.5 cm).

We let this ferment at ambient temperature for three months and didn't mix it. We contained it with an air lock as you would for alcohol fermentation.

After a few months it was really boozy. As with any koji ferment, the sugars were broken down and the natural fermentation of sugar into alcohol had occurred. It only made sense to us to dehydrate the seed-grain mixture to dry the solids, concentrate the flavor, and evaporate off the alcohol. The results were awesome. The sunflower-barley combination can be best described as a savory, salty, and slightly tart granola. The notes of lactic acid from the fermentation really made it sing. We gave some to our friend chef James Wayman of the Oyster Club in Mystic, Connecticut, who mixed it in with a Mexican-inspired rub for a chicken confit. Aside from the umami hit, the touch of acidity cut through the fat so you could really taste the chicken.

Of course, this variation is not a douchi, a soy sauce, or even a miso. It's an amino-driven make that has taken pieces of that fundamental understanding to create an interesting amino *salt*, a derivation that demonstrates how traditional failures in the process can transform into something wonderful. This is just one example of what we've been able to do when fermenting with koji.

## Returning to Natural Fermentation

The key element to everything in this book is using a koji starter culture, the spores of various species of *Aspergillus*, primarily *A. oryzae*. Throughout the entire book, we haven't encouraged you to begin a wild fermentation to harness the molds. That's because we haven't gone down that rabbit hole ourselves. We've been so busy investigating the ins and outs of koji, we haven't had the bandwidth. Also, it is one of the more difficult fermentation processes to become comfortable with. There's always the concern that you may by chance nurture the wrong microbes and become ill from the make, especially in locales far removed from where koji evolved—such as the ones we live in. We are here to tell you, however, that it *can* be done under the right conditions following established processes. It's something that's been practiced for thousands of years by the Korean people—see the "Meju" sidebar—and it's at the core of their condiments. It can't be that much more difficult than making koji.

## MEJU

### John Hutt and Irene Yoo

*Since we lacked the experience to present this topic, we turned to our friends John Hutt, executive chef of MOFAD (the Museum of Food and Drink) in Brooklyn, New York, and Irene Yoo, the photography manager at Food Network. Both of them have an in-depth understanding of Asian cuisine and are passionate researchers when it comes to understanding everything about a subject that intrigues them. It just so happened that they had a batch of meju started when we got in touch with them to research the process. They were more than willing to fill in the blanks on what they had learned and intensively researched. In this sidebar, they share what meju is, how to make it safely, and what to do with it.*

Meju is a soybean-based wild fermentation starter for Korean *jang*, an amino paste. It is simply cooked soybeans that are wild-fermented to become rich in *Aspergillus oryzae* and *Bacillus subtilis*. Meju can be best understood in the Korean framework, as that is where it is predominantly made. When dealing with fermentation, especially wild fermentations like meju, the location where it's made becomes an ingredient itself. The environment that fermentation occurs in lends different types of bacteria, wild yeasts, and mold—and the specific fermentation techniques used will allow for good bacteria to thrive and prevent pathogenic and spoilage bacteria from ruining the product. This is especially true in Korea, where the concept of *son-mat* (hand taste)—the idea that your literal hands add flavor—is prevalent in all fermentation and cooking, particularly kimchis and jangs.

Jangs are the basis of Korean cooking. These soybean pastes and sauces are the building blocks of many Korean stews and dishes. The three main jangs we will refer to here are:

**Ganjang.** Specifically *guk-ganjang* (soup soy sauce) or *jip-ganjang* (home soy sauce), a type of Korean soy sauce made with meju and a saltwater brine. It is used as the base of Korean soups and *namuls* (vegetable dishes).

**Dwenjang.** This fermented soybean paste made with meju and saltwater brine is the by-product of ganjang making. It is used as the base of Korean *jjigaes* (stews) and marinades.

**Gochujang.** A fermented red chili paste made with chili powder, meju, glutinous rice, and *yeotgireum* (barley malt powder).

Meju is very similar to its ancestor, the Chinese qu, and has been made in Korea for centuries. While the origins of meju can be traced back to qu, there was never a school of fermentation cultures on the mainland sending out professors and masters to ensure that qu production is up to the Chinese standard.[8] Rather, people were introduced to a technique and played with it, adjusted it, and grew it into a new technique. When we say "Korean-style soy sauce" has been made for hundreds of years, we know that it is not one person but a whole group of makers. Fermentation, and to a larger extent food production, is an oral and physical tradition passed down through generations as techniques evolve. The techniques of traditional meju were passed down to the present time by Korean housewives.

Meju as it is referred to today means a hybrid wild-fermented starter culture based simply on soybeans, which becomes inoculated with *Bacillus subtilis*, *Aspergillus oryzae*, and myriad ever-changing zygotes, bacteria, and fungi depending on the weather, the air, the hands and sweat of the cook, and the location.[9] Meju differs more significantly from miso in that meju is usually purely soybeans that are wild-inoculated, while miso is soybeans (or grains and cereals) specifically inoculated with *A. oryzae* (koji), and often mixed with wheat or other ingredients. Meju also relies heavily on fermentation through the air, while miso relies on anaerobic fermentation. *Wild fermentation* loosely means fermentation without a starter. These are the easiest ferments to make and include makes such as sourdough starter, wine, and meju. Often these wild starters serve as the basis for other schools of ferments.

A starter can be a jar used for a few cycles to make yogurt, a room full of country hams, a box of rice husks, hay, or pandan leaves, a breeze next to a rice field, and even honey. As the definition gets more abstract, it becomes easier to realize that what we are discussing is a minor adjustment of the natural course of things. What seems to be a totally obscure technique like ganjang production becomes a logical course of next steps when going on to make dwenjang and is as inevitable in the course of Korean cooking as grilling is.

### MEJU MAKING OVERVIEW

Making a wild-fermented starter like meju is as simple as a sourdough starter, just with a longer time component and different ingredients. To make meju there are four steps: Make soybean bricks, dry the bricks, ferment the bricks, and dry the bricks again. Once you have made meju, you can submerge it in salty water for a few months, then separate the solids and liquids into dwenjang and ganjang respectively. The other popular method is to grind meju into powder and add it to soybeans and chilies to make gochujang.

Meju for gochujang is made using soybeans mixed with rice, barley, or wheat. The ratio is 6:4 soybean to wheat, or 5:2 soybean to glutinous rice.[10] Meju made for gochujang is not typically kept for a long time. This process is actually closer to the Chinese technique for making solid-state soy sauce, the ancestor to ganjang.

### MAKE YOUR OWN MEJU

First, begin with 2.5 kilograms (5.5 pounds) of dried soybeans. Sort through the beans and remove any pebbles or shriveled brown beans, keeping only the whole, unblemished yellow and white ones. Rinse the beans underwater to clean. Soak the beans overnight in three times the amount of water to beans, at around 68°F (20°C). Drain the beans and either boil them for 4 hours, or steam them for 2 hours.[11] Cool the beans to 105°F (40°C), roughly mash them, and then form them into bricks with a dimension of 3 × 4 × 8 inches (8 × 10 × 20 cm).

Allow the bricks to dry. To accelerate the drying, place them either in a drying area at about 105°F (40°C), or on top of a tray on a radiator, or in the boiler room. These locations have the added bonus of jump-starting the growth of bacteria and yeasts in and around the brick, but it is by no means necessary. Once the bricks have dried enough to handle, they must be hung for 40 to 80 days in an environment that averages no more than around 68°F (20°C).[12] During the first week or so, the brick will smell pungently stinky (don't worry about this, it's totally normal). After the initial drying and fermentation, the dominant species of growth on the meju is *Bacillus subtilis*, and in order to complete the fermentation you should inoculate with *Aspergillus*.

Generally *Aspergillus* is added by placing the meju in a box or other container with dried rice stalks, but if those are not available you can use dried hay, straw, hibiscus, or even pandan leaves as is common in Indonesia.[13] Once the meju is in the container with the straw, place the box in a warm environment such as on top of an electric blanket (which approximates the Korean heated floors called *ondol*), on a heater, or in a boiler room. The ideal temperature is about 100 to 105°F (37–40°C), in an environment with high humidity. Allow the blocks to ferment in the box for 14 to 30 days.

Remove from the box and allow to dry a further 14 to 30 days or until needed. The blocks will have fungal growth on them, predominantly white to green (which is *Aspergillus*). At this point you have made meju, which you can use as a starter for ganjang, dwenjang, or any other amino paste you wish to create. Finished mejus can be stored almost indefinitely.

### WHAT IS GROWING IN MY MEJU?

The meju block, whatever the variant, is a breeding ground for molds and bacteria that are vital to its existence, and its use, and this does not mean it is unsafe to eat. Before using the meju to make ganjang or gochujang, you must clean the outside to remove most of the fungal blooms, some with more vigor than others. Anything black, gray to black, green, or orange is best scrubbed off thoroughly. You can also remove the affected layer prior to use. Here's a list of known microbes that grow in and on meju:

**Fungi:** *Aspergillus oryzae, Mucor abundans, M. griseocyanus, M. mucedo, Murocales* spp., *Penicillium kaupscinskii, P. lanosum, Rhizopus chinensis, R. japonicus, R. nigricans, R. oryzae*, and more.[14]

**Yeasts:** *Rhodotorula flava, Saccharomyces coreanus, S. rouxii*, and *Torulopsis datria.*

**Bacteria:** *Bacillus pumilus, B. subtilis*, and *Staphylococcus aureus.*[15]

### HOW TO USE YOUR MEJU

Meju is most commonly used to produce ganjang and dwenjang by soaking in a saltwater brine for several months. These are then strained and separated and further fermented separately, which ensures the longevity of each product. If they continue to be stored together, this is called *toejang* and does not have a very long life span.

To make ganjang, wash the meju blocks in cold water, scrubbing lightly to remove fungi and molds from the surface of blocks. Dry the blocks for a day, turning so all sides are exposed. Place meju blocks in an earthenware crock and cover with a saltwater brine (10 percent salinity), at a ratio of 1:1:3 meju to salt to water. Charcoal, jujubes, and dried Korean peppers can be added to taste or not at all if you choose. Cover and let ferment for 2 to 3 months. The resulting liquid should be drained and boiled for 30 minutes, and then it is ready. It can be further fermented and aged allowing the taste to go from fish sauce funky to mellow saltiness over the course of several years. If any mold mycelium forms on the top as it ferments and ages, it can be easily scraped off.

Dwenjang is the fermented soybean (the leftover meju) reserved from the ganjang drain. It can be mashed into a paste, placed in the earthenware crock, covered with salt, and further fermented for 6 months. Ganjang can be mixed in to keep the paste's texture intact.

### OTHER MEJUS

A common alternative meju is meju made for *gochujang*. This is prepared using soybeans mixed with rice, barley, or wheat. The ratio is 6:4 soybeans to wheat, or 5:2 soybeans to glutinous rice. Meju made for gochujang should not be stored for a long time (use it to make gochujang). This meju is powdered and mixed with malted barley, sweet rice flour, and chili powder, after which it is further fermented. There is also an "improved meju," which is made with soybeans specifically mixed with *Aspergillus oryzae* to control and shorten fermentation time.

There are many variants of meju. While soybeans are the most popular ingredient to use, meju is also made with sweet potatoes, yams, broad beans, and wheat. Essentially anything that has a high starch and protein content is usable. Each variant has its own traits. When making *teinmenjiang* (or *chunjang* in Korean), for example, bread wrapped in melon skin is used, resulting in a sweeter paste. If you make a meju block with rice and barley, this is called *nuruk* and is commonly used for starting alcohols such as rice wine.

As you can see the complexity of meju and its derived makes is essentially limitless. Your creativity can drive you to make myriad types of amino pastes, each with its own identifying and unique story to tell.

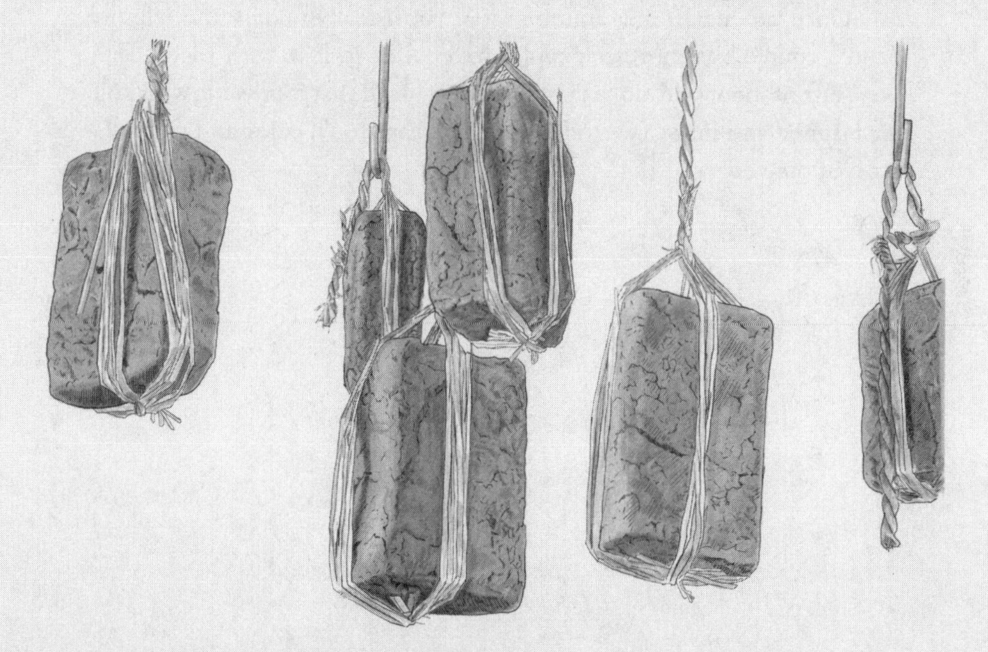

Traditional meju blocks hanging. *Illustration by Max Hull.*

## Don't Be Daunted

We feel that amino pastes are one of the easiest gateways into understanding the incredible flavor power of a koji-based make. One challenge for those who aren't sitting on a few ready to go in their pantries is that they do take a while. Our recommendation is to head to your nearest Asian market to check out everything they have to offer. The variety is amazing and can be overwhelming. As with anything that you're trying for the first time, do a little research and ask family and friends who are familiar. If all else fails, adventurously pick a few that seem to have what you're looking for and start using it in a savory application in place of salt. You can think of amino pastes as you would a pesto in application; there really isn't a wrong way to use them. A few straightforward applications that work well include mixing somewhere between a teaspoon and a tablespoon of amino paste into a basic soup base or pan sauce, or coating vegetables ready to roast.

Once you start using amino pastes, you'll really understand why we (and a couple billion other people) have fallen in love with them. They were one of the applications that hooked us on all that is possible with koji. Be inspired and make a custom condiment that you'll enjoy and will tell a story about you.

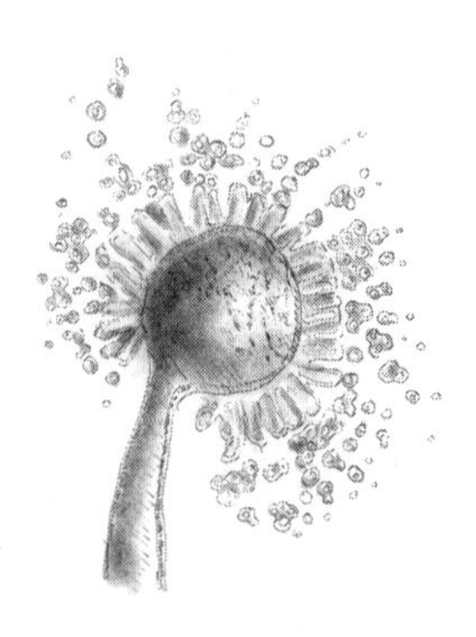

CHAPTER 8

# Amino Sauces

Amino sauces aren't very different, functionally, from amino pastes. However, comparing two ramen styles, miso and shoyu, is a great way to highlight those differences. Or consider the reasons why soy sauce is commonly matched with steak, while miso is paired with fish. Each has distinct characteristics even though both have a koji-driven umami core.

The primary difference between an amino paste and an amino sauce is their water content. The higher liquid concentration in an amino sauce makes the koji enzymes more accessible to starches and proteins to assist in a more complete conversion into sugars and amino acids. This is especially important if you're creating an amino sauce using meat or fish scraps, as you'll need to get into all the tiny nooks and crannies created where the flesh and other bits were removed from the bones.

Amino sauces provide a delicious accent to coat, drizzle on top of, and dip food into, making it more tasty. By definition, sauces add moistness and flavor to foods. In one of the most important works about gastronomy ever written, *On Food and Cooking*, author Harold McGee offers a wonderful description of sauces at the beginning of the chapter dedicated to the subject.

> *Sauces are liquids that accompany the primary ingredient in a dish. Their purpose is to enhance the flavor of that ingredient . . . either by deepening and broadening its own intrinsic flavor, or by providing a contrast or complement to it. . . . [S]auces [are] the ultimate composed flavorings. The cook conceives and prepares sauces for particular dishes, and can give them any flavor.*[1]

As part of this section, McGee categorizes soy sauce and fish sauce as condiments, and we concur. He adds, however, that sauces sometimes include condiments—and that's the connection we want to make here.

Most of the savory sauces and condiments we know and love have amino acids in them. When we think of our favorite sauces—tomato, cream, certainly anything involving cheese—all of them contain strong concentrations of glutamic acid. America's popular condiments mayonnaise and salsa are also driven by the same umami flavor agent. Why do we love amino acids so much? Why do they make our mouths water? Why does eating them feel so satisfying? It all comes down to the fact that they're an indicator of protein, a rich source of nutrition. It makes sense that we are

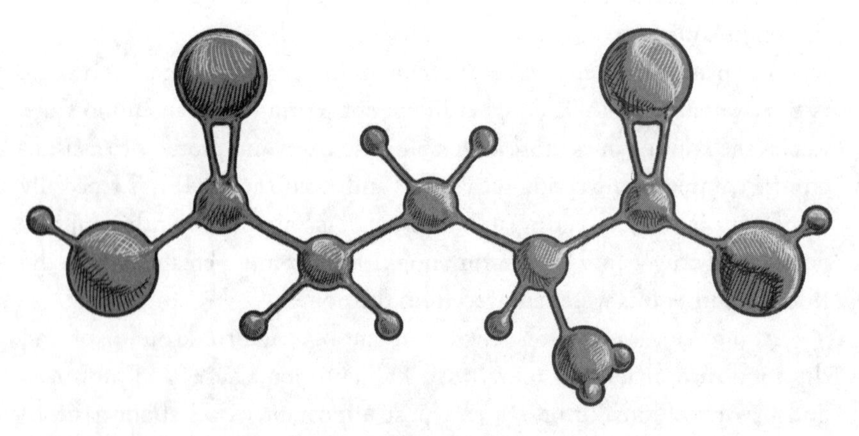

Glutamic acid molecule. *Illustration by Max Hull.*

biologically wired to detect amino acids, since savory foods high in protein give us great amounts of energy and help us to feel full longer than those high in sugar.

However, this basic need can be taken advantage of. The mass-production food industry, driven by profit margins and greed, leverages our innate desires by taking simple base recipes and then bumping them up with MSG or equivalent glutamates. Did you ever wonder why Japanese mayonnaise tastes so damn good? It's enhanced with MSG! Of course, so is the all-American salad dressing we know and love: ranch dressing. Bottom line, it's all a matter of degree and how far someone chooses to take it. And who doesn't love a little MSG every now and then? We surely do.

## You Are in Control

When it comes to making amino sauces, you control the flavor power, which is driven by the ratio of protein to koji (ultimately, protein to protease enzyme), which in turn yields amino acids. The traditional amino sauces that we typically think of—soy sauce and fish sauce—are loaded with umami. However, you don't have to go crazy with protein loading to get there. Shio koji alone can essentially be considered an amino sauce. As we've mentioned before, the inherent proteins in the grain provide enough of a touch of full flavor that "less is more" literally makes sense. Whatever ingredient you have on hand can be used to leverage the enzymatic power of koji and be made more delicious. Keep in mind that going all-out on the umami isn't always the right answer. Finesse is key.

As noted previously, the key characteristic of an amino sauce is its high water content. With all the ingredients in a brine solution, it's easy for enzymes and microbes to move around. They readily come in contact with the other ingredients in the sauce and break them down to create intense depth of flavor. We have identified three driving factors to pay attention to when making these sauces: maximizing flavor via autolytic conversion; the potential to use "waste," such as hard indigestible and "inedible" bits; and ensuring proper food safety.

It's simple to make a delicious condiment from pretty much anything with protein content. Now "just add water" has an entirely new meaning.

## UMAMI DUST, OR MSG?

### Coral Lee

Even though the myth that MSG causes headaches and other health problems has been debunked numerous times, visibly and again recently by Jennifer LeMesurier (writing and rhetoric professor at Colgate University), David Chang (owner of Momofuku Restaurant Group, host of *Ugly Delicious*), and the Japanese food brand Ajinomoto (producer of those panda MSG-shakers), still the negative stereotype and baseless fear persists. Why?

In *Uptaking Race*, Jennifer argues that "the mistrust of Chinese immigrants manifested itself in anti-Chinese food rhetoric," and "genre uptake—the process of information selection and translation from medical discussion to popular news—facilitated this prejudiced understanding."[2] Underlying racism and xenophobia manifested in the rhetoric surrounding and consumption of goods—comestible and not. It's easier to complain of a headache than it is to admit you're afraid of Chinese people.

This runs counter to our universal fascination with and celebration of umami. Umami is aloof, mysterious, sexy; MSG is *not*. But smart copy and embodied rhetoric aside, the detection of this elusive fifth flavor is merely the tasting of amino acids ("free glutamates") or an indication of protein; MSG is the isolated, most stable form of amino acids (that is, glutamates, no longer free). As you're now very aware, as koji works its way through your chosen substrate, it produces enzymes, one of which (protease) is responsible for breaking down proteins into individual, readily accessible amino acids. The only thing that differs among these three is the method of unlocking, or setting free, glutamates for your palate.

But as with most things in life, too much of good thing perverts said good thing. While you now may be tempted to shake MSG into a cocktail, umami-dust your burger, or koji your everything—encouraged by how natural these enhancers are—remember what you are tasting is the presence of free amino acids, not complete, actual proteins. So throw some greens in with your choice of umami to round it all out.

In the case of shoyu, the Japanese version of soy sauce, the complexity is best described by an excerpt from "The History of Soy Sauce, Shoyu and Tamari" by William Shurtleff and Akiko Aoyagi:

> *In shoyu, the* Aspergillus oryzae *molds grow directly on the soybeans (and wheat) during the koji fermentation and their enzymes begin to digest the koji substrates, then continue to digest the soybeans (and wheat) during the subsequent brine fermentation. This* in vivo *and* in vitro *extended hydrolysis leads to the formation of complex metabolic compounds, a higher degree of protein hydrolysis and liquefaction, and the production of a richer and stronger flavor in shoyu than in miso.*[3]

Compare this brine delivery system with a thick blended paste like miso and you will see that liquid has the distinct advantage of being able to easily surround and fill all the nooks and crannies of any solids. This means that you can feel free to use the indigestible odds and ends of a product to extract flavor from it. We're talking about bones, shells, seed hulls, rinds, skins, and so forth. This is the very reason why fish sauce contains the entire fish: bones, guts, and all.

Making a stock employs the same concept. Extracting the flavor from the combination of flesh and bones simply by simmering them in water isn't really different from making an amino sauce. Interestingly enough, to add depth to a stock, you can roast the bones ahead of cooking them in water to extract the flavor. That's not far from the idea of toasting wheat ahead of making koji for an amino sauce. Functionally, the difference between making a stock and making an amino sauce is that with the former you use heat to accomplish the flavor extraction, and with the latter you're leveraging enzymes and fermentation to do the work.

On the surface, the relationship between the two main, traditional sauces—soy and fish—seems to be a stretch, but if you dig a little deeper, the makes are strikingly close. Aside from their shared popularity as condiments, they both represent a time when food had to be preserved and shelf-stable for long periods of time. Salt brining is easy to execute when a huge seasonal peak harvest occurs, so a simple solution was just to stick food in vessels with salt and water, then wait. It doesn't get easier than that.

## BASE FISH SAUCE RATIO

### Sam Jett

One thing to remember when making fish sauce is that the quality of the main ingredient is paramount. The fish needs to be impeccable, very fresh (if not still swimming) when you get it. The higher-fat-content fish will create much better fish sauce (although the fat must be separated before bottling as it is not good at all). Try to keep the fish guts in; if they are not present, then you cannot do the traditional method of high salt and a long fermentation time. If there are no guts, don't fret: Koji will provide the enzymes necessary for the autolysis to occur.

In table 8.1 I have laid out the salt and koji ratios I use when making fish sauce. If I am pushing for a quick turnaround, then I will reduce the salt and cook at 140°F (60°C). I can pull an intensely flavored amino sauce in 2 months at this $a_w$ (water activity; see appendix B) and temperature. I normally scale the water activity and temperature down as I increase the ferment time. Anytime I am going for a yearlong process, I will just do this in a traditional fashion and have a high salt content.

To process the fish, don't rinse or clean it in any way. With gloves and a cleaver, chop the fish into 2-inch (5-cm) chunks and mix with the other ingredients. If you are using smaller fish like anchovy or sardines, mix with the salt and run through the meat grinder before mixing with the koji and water. Place in an airtight crock with an air lock to help prevent loss through evaporation. Store in your chamber (or on a hot roof) for as long as desired.

Fish for an amino sauce. *Illustration by Max Hull.*

**TABLE 8.1.** Fish Sauce Ratios and Times

| | 2 Months | 6 Months | 1 Year + |
|---|---|---|---|
| FISH | 100% | 100% | 100% |
| SALT | 8% | 12% | 16+% |
| KOJI | 20% | 20% | None |
| WATER | 10% | None | None |
| TEMP | 140°F (60°C) | 113°F (45°C) | 84°F (29°C) |

But there is a distinct difference between adding brine to some ingredients and making an amino sauce: enzymes. Without these, you'd just have salty fermented food, which, while good, isn't nearly as *delicious*. The koji in soy sauce provides the enzymes that break the plant proteins down into amino acids. Traditional fish sauce, composed solely of fish, isn't made with koji; however, it's also driven by enzymes—those found in fish digestive tracts. There's something beautiful about a basic process that uses an entire fish to create a delicious sauce.

Of course, food safety is always on our minds when we are preserving food with fermentation. It is especially a concern when meat and fish products are used with bones intact. The reason for this is that the cavities vacated by the bones have the potential to allow pathogens to thrive. This fortunately becomes highly unlikely when all of the ingredients are broken up well and submerged in a brine solution.

## The Basics of Soy Sauce

The typical process for making a soy sauce utilizes a 1:1 ratio of toasted, cracked wheat and cooked dry soybeans to make a koji, then adds that mash to an equal amount of water. The amount of salt varies depending upon the process. Fifteen percent salt by weight is an easy number to remember that falls between the extremes. When you make a soy sauce, the mash follows a mixing schedule based on fermentation activity influenced by temperature: once or twice a day for the first month followed by once every one to two weeks works well for small batch production.[4] "Mixing of the mash introduces fresh air into the mixture, allowing the growth of yeast or other beneficial microorganisms and inhibition of the growth of undesirable anaerobic bacteria. At the same time, it helps evaporation of unwanted carbon dioxide and hydrogen sulfide, oxidative coloration of the product, and homogeneous mixing of the mash."[5] The mash ferments for a year until it's harvested. When it is ready, the solids are strained out and the liquid is cooked to a desired concentration. In the process, the soy sauce is pasteurized.

Now that you understand the specifics of the soy sauce process, it's easy to see how you can interchange ingredients in the same way we described for amino pastes. The simplest way to start is using the ratio of 1 part koji to 1

## BURNED BREAD AMINO SAUCE

### Kevin Fink

*Kevin Fink is the chef and owner of the restaurant Emmer & Rye in Austin, Texas. Kevin is one of the most dynamic and talented chefs in America right now, as evidenced by him being named a* Food & Wine *Magazine Best New Chef and a James Beard finalist for Best Chef: Southwest. One of the things that makes Kevin and his restaurants so special is the larder that he maintains. At Emmer & Rye you will see an extensive collection of all things fermented, from sodas to amino pastes. Kevin and Jeremy forged a fast friendship and bonded over their mutual love for koji a number of years back when Jeremy went down to Emmer & Rye to work with Kevin in his larder. Since then Kevin and his amazingly talented team of chefs have continued to push the envelope of what cuisine from the foothills of Texas is. One of our favorite makes to come out of the larder at Emmer & Rye is this Burned Bread Amino Sauce.*

At Emmer & Rye we always have a naturally leavened bread dish on the menu. We showcase different grains by baking and serving fresh bread every day. We always end up with some extra bread, so we have come up with myriad ways to use it. We've used it for bread crumbs, fried it for *panzanella*, turned it into toast with Ikra and vinegar—and sometimes we just eat it. Burned Bread Amino Sauce is by far my favorite. It represents the third time the grain has been fermented—the first two are when it is made into a starter and then bulk-fermented in dough form. When the bread is finally introduced to koji, its last transformation begins. Proteins and carbohydrates are broken down, creating a deeper presence of umami, while the increased sweetness rounds out the bitter, highly caramelized flavor of the bread. What you end up with is a rich sauce that can stand by itself or be used to enhance other products.

| | |
|---|---|
| **1,500 g (3.3 pounds) stale bread**<br>**3,000 g (6.6 pounds) water**<br>**300 g (10.5 ounces) barley koji (*Aspergillus oryzae* suited for shoyu production)**<br>**330 g (11.6 ounces) salt** | Grind the stale bread into crumbs and then spread them onto a roasting tray lined with a Silpat about 0.5 inch (6 mm) thick. Roast the crumbs at 325°F (163°C) for 45 to 50 minutes or until they have achieved a dark brown color. Allow the bread crumbs to cool to room temperature and then combine them with the rest of the ingredients. Keep under Cryovac for 4 to 6 months, or keep in a jar under a crock weight. Strain and use. |

part protein and 2 parts water. Blend it all together as the base of your amino sauce. We typically add 7 to 10 percent salt, since we don't typically wait as long as a year to enjoy our sauces. Another thing to consider is the flexibility of the water content. If you want to decrease the water to 1 part instead of 2, you end up with a more concentrated amino sauce that's closer to a tamari.

Vegetable protein can be whatever you want it to be. Of course, it's optimal for the vegetable to have as high a concentration of protein as possible with low carbohydrates and fats. The solids of nuts and seeds after they've been pressed for oil are ideal. By default, they represent a defatted and high-protein base that would have otherwise gone to waste. That's a pretty wonderful thing.

When an amino sauce is at a stage where you like its flavor, follow the same process as standard soy sauce making: Strain out the solids and bring the liquid to a simmer to stop further fermentation. The remaining solids can be used in a similar fashion to amino paste but with less flavor. We also like to dry the solids out for a seasoning salt or to ferment fresh root vegetables, such as beets, to make pickles.

We've all been guilty from time to time of not following prescribed practices with our ferments for whatever reason. If you forget to stir your amino sauce every once in a while or keep it covered for an extended period, you may note a level of unwanted alcohol that builds up. Even though it's not ideal, you can simmer this to cook off the unwanted flavor and in most cases the sauce will still turn out to be a pretty good product.

## Elevate Your Scraps

Although amino sauces can be made from the best of grains and legumes, their beauty lies in their ability to also incorporate scraps, otherwise considered food waste, to make a delicious condiment. This has economic benefits not only for chefs looking out for their restaurant's bottom line, but also for home cooks. Less waste generated also means fewer resources we have to invest into finding a place to put the waste. In today's disposable culture this benefit should wholeheartedly resound with every one of us. Now you can start to see what we meant by soy sauce being the tip of the koji iceberg.

### REJUVELAC AMINO SAUCE

**Sean Doherty**

*Sean Doherty is a baker in Portland, Maine, by day and a fermentation mastermind the rest of his waking hours. You see folks riffing on shio koji with all sorts of different grain bases these days, but nothing has blown our minds more than Sean's decision to use rejuvelac, a probiotic drink made by soaking sprouted grains, as the core for an amino sauce. We had the opportunity to taste a batch at seven months and it was crazy delicious. It made us really think about all the amino sauces you could make with less protein, and with other levels of complexity.*

**4 cups (1 L) of any whole grain**
**1 widemouthed gallon (4 L) jar**
**Cheesecloth (enough to cover the jar mouth with a few layers)**
**1 rubber band**

Soak the whole grains in a jar of water for between 12 and 24 hours at room temperature. Drain and discard the water. Place the jar upside down in a bowl or other container large enough to tilt the sprouting jar, so the grains don't touch the bottom. Rinse the grains with fresh water at least twice a day, ideally morning and night, to keep them from drying or worse, molding. When you notice little tails starting to grow, usually between 2 or 3 days, they are ready. Rinse the sprouts one last time and then add them back to the jar. Fill the jar with water and loosely cover to keep away dust and flies for about 48 hours. Now the rejuvelac is ready. Pour off the liquid and refrigerate, saving the sprouts. Steam the sprouts for 25 to 30 minutes, but keep the structure of the grain in mind as we've previously covered. Make koji using the cooled, cooked sprouts as the base following the same process as you would for rice. When the koji is ready, recombine the sprouted koji with the saved liquid (brought to room temperature) plus 5 percent salt brine in a ratio of 3:1 by volume. Place mixture in a non-reactive (glass or stainless steel) container, loosely covered. Keep it in a warm spot that's somewhere between 80 and 90°F (26–32°C) for 3 to 4 weeks. Stir and taste daily for the first week and then once a week afterward. I refrigerated mine after 2 months; after 9 months it tasted quite delicious. A small amount of alcohol was produced during the fermentation that seemed to balance everything out nicely.

Note: Kept longer at room temperature, the sauce gets funkier and even more sour due to different strains of *Lactobacillus* taking over, mainly *L. plantarum*.

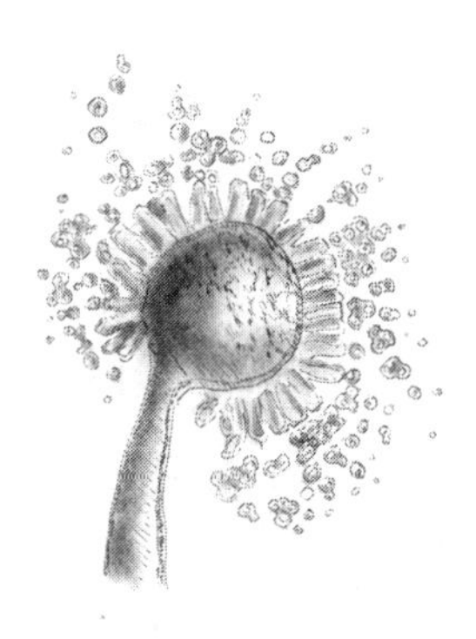

CHAPTER 9

# Alcohol and Vinegar

*Making booze is easy—keeping it around is the hard part.* This sentiment can be interpreted in many ways: from the ability of alcohol to quickly evaporate, to it being a delight to consume, to it easily being fermented into vinegar. Whatever way you choose to read it, alcohol is a substance of great significance for most of humanity. We use alcoholic beverages to help us celebrate everything in our lives from the birth of a child to the death of an old friend. They have played important roles in religious lore and dogma from Bacchus to Jesus, and are used to cement and reaffirm many of our individual faiths. Governments have suppressed it and oppressed its users only to be beaten back to the point of allowing it flow freely again, mostly.

Earlier in this book we discussed the historical importance of alcoholic beverages as being necessary for humans as we spread to distant lands. We might not know exactly when the brewing of alcoholic beverages became widespread, but most historians and scientists agree that it was sometime between nine thousand and twelve thousand years ago, essentially coinciding with the dawn of agriculture. Could the domestication of various cereal crops and fruits have been primarily driven by our love of booze? Sure, that could have happened, but we'll never know with 100 percent certainty. What we do know is that this ancient love affair only grows stronger as time goes on, leading to countless styles of alcoholic beverages. And koji can be used to make them all.

In the late 1800s a Japanese man by the name of Jokichi Takamine proved this to the world by using koji to create whiskey. While Takamine wasn't successful in his ventures for various reasons, he showed us that koji's unique ability to rapidly saccharify starches could be exploited and worked into nearly any style of alcoholic beverage. Takamine's spirit is being evoked around the world today by such talented brewers as Lars Williams at Empirical Spirits in Copenhagen, Denmark, and Brian Benchek and Jason Kallicragas at the Bottlehouse Brewery & Mead Hall in Cleveland, Ohio. They are pushing the boundaries of how koji can be used to create some of the most delicious alcoholic beverages ever.

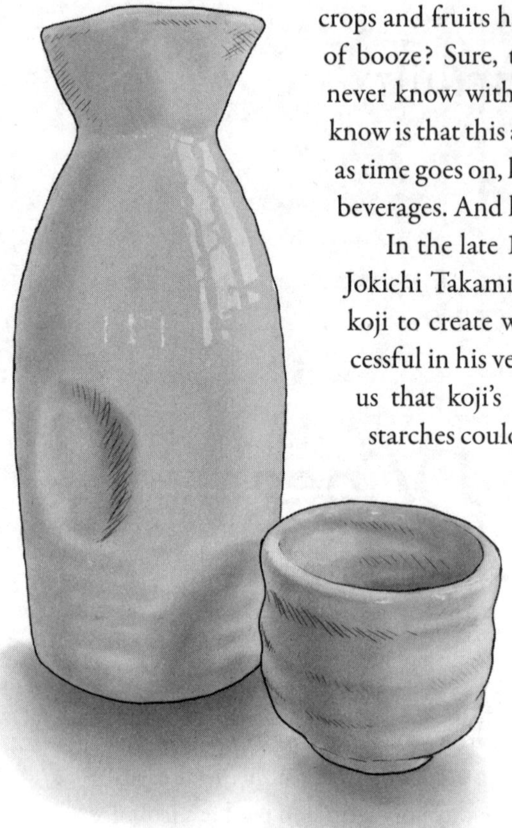

Sake set. *Illustration by Max Hull.*

## Depth and Breadth of Preexisting Alcohol

Our friend John Hutt (who previously shared expertise on meju) is a koji-based-alcohol obsessive. He points out that to catalog the various types of alcohol and distillates made with koji would take several lifetimes. The intricacies associated with making alcoholic beverages with (or without) koji are so minute and potentially overwhelming that this subject could easily have its own book of comparable or grander length to this one. Still, we want to cover the fundamentals in a way that will allow you

## KOJI BEER BREWING INGENUITY

**Brian Benchek**

*Brian Benchek's work integrating koji into alcoholic beverages that are traditionally made with malted grains is particularly fascinating. His brewery, the Bottlehouse Brewery in Cleveland Heights, Ohio, has always focused on small-batch, seasonal production of a wide variety of beers and mead. He and head brewer Jason Kallicragas focus on integrating the terroir of the Midwest and Great Lakes into beverages that can showcase the biodiversity of this area by using not only local ingredients but also local and wild microbes. Brian's brews are a fantastic example of not only what koji can do when used in the brewing of beer but also of what happens when European-style beverages are made with Asian ingredients using American ingenuity.*

The American brewing industry is young. When compared with the rich and extensive brewing culture of many European countries, we have just received our driver's license and are heading out on the open road without a map. And while we may not know which direction to go or which road leads to a dead end, we know that there is plenty to be discovered along the way.

Although traditional European brewing methodologies provided a starting point for the great American craft beer journey, the creative and curious nature of the American brewer, free from the burdens of tradition, has, over the last two decades, altered the global brewing landscape. This creative evolution is driven by a continual search for new processes, new ingredients, and new methodologies to incorporate into the brewing process. Oftentimes brewers look to other artisanal producers when seeking inspiration. The utilization of koji in the brewing process is an exciting example of this.

Whether found in beer, bourbon, vodka, or sake, ethanol production begins with the same process, converting polysaccharides (starch) to monosaccharides (glucose). An enzyme known as *amylase* is the primary driver of this conversion. Without this enzymatic conversion, starch molecules would be too large for yeast cells to "digest" and we would live in a world without starch-derived alcohol. Brewers and distillers rely on malted barley or malted wheat as a source of the necessary amylase enzyme. (In contrast, sake producers find their source of amylase in the growth of koji.)

While each source (malted barley or koji) produces the amylase enzyme necessary for the starch conversion, the

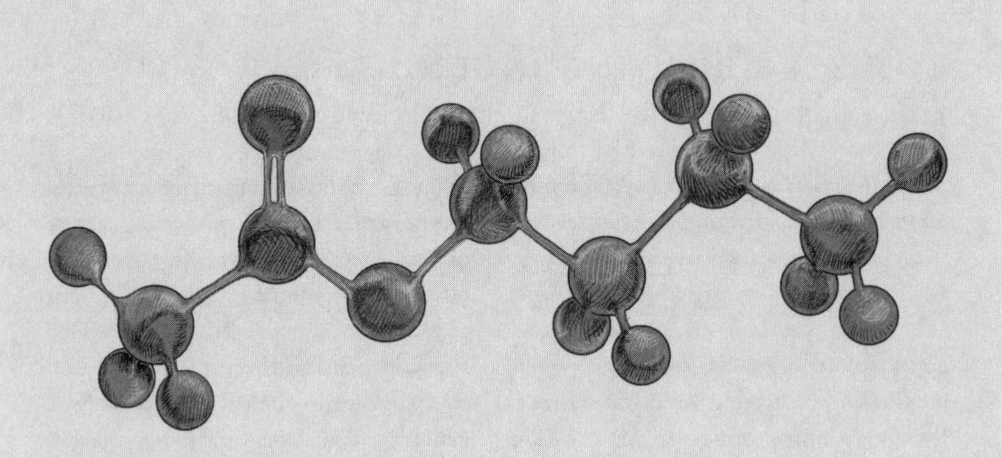

Ester molecule. *Illustration by Max Hull.*

difference in the specific types of amylase (alpha- or beta-amylase) produced and the unique by-products produced by the individual processes contribute to the overall flavor, aroma, and mouthfeel of the final product. It is within these differences that brewers see the potential for incorporating koji into the beer making process.

Koji can be used by brewers in several ways. Since koji cultured on rice produces large amounts of alpha-amylase, brewers can replace a portion of their grain with koji-cultured rice. By adding the koji-cultured rice directly to the mash, the level of alpha-amylase in the mash increases dramatically. The effect is threefold: First, brewers will see a decrease in the time required to convert the polysaccharides in the barley into fermentable sugars. This has the practical benefit of decreasing the length of a brew day. Second, the increased alpha-amylase in the mash creates a more fermentable wort, which means less residual sugar and in turn a lighter body in the final beer. But perhaps the most exciting aspect of incorporating koji-cultured rice into the grain bill is the addition of the fruity esters created by the koji when breaking down the starch in the rice medium on which it grows.

Unlike the amylase conversion on barley or wheat starch, when koji converts the starches in rice to glucose it also produces a variety of fruity esters. These esters tend to have floral flavors and aromas similar to guava or tropical fruit. These esters are responsible for the subtle fruit flavors and aromas found in sake. However, the delicate nature of the fruity esters produced by koji can easily be destroyed when the wort is boiled. In order to preserve these flavor and aroma compounds, brewers have looked for other ways to incorporate the esters produced by koji post-wort-boil. To do this brewers have collected the mirin produced by the koji-cultured rice and added it back to the beer when it is nearing the completion of the fermentation process. The mirin is rich in fruity esters, which enhance hops varieties known for their tropical characteristics, as well as the fruity profile of many yeast strains, specifically Belgian strains.

The downside to using mirin as the sole source for the fruity esters created by koji is the volume needed to have a noticeable impact on several hundred gallons of finished beer. At the scale a small brewery operates, there simply isn't enough mirin created during the koji culturing process to produce the results brewers are looking for. One solution brewers have implemented is to make large batches of amazake, strain the rice solids out, and add the amazake directly to the fermenting beer. This process allows brewers to add the necessary amount of fruity koji esters without having to culture the large amounts of rice needed to collect the needed volume of mirin.

To brewers who constantly seek to extract more and more fruit aroma and flavor from the hops and yeast used in the brewing process, the appeal of incorporating koji to enhance these characteristics of the final beer cannot be overstated. However, the use of koji in beer brewing is still in the experimental phase. Altering the traditional methods and processes of using koji mold to fit the needs of beer brewers will require time and a willingness to think outside the proverbial box, a philosophy American craft brewers have eagerly embraced.

to further explore the endless possibilities that koji-based alcohols and vinegars will afford you.

Chef Hutt also reminds us that the fermentation of alcohol, with or without any starter culture, is so old, and each part of the process matters so much, that any minute change results in a different product. There are two main ways to approach brewing alchohol that strem from this: as an encyclopedist, and as a practitioner. The encyclopedist quantifies and classifies beverages based on region, material, method, age, color, filtration, additives, and so on. This is important because if we had no way to discuss the massive numbers of beverages, techniques, or starters, we would be starting from the beginning each time. The practitioner, on the other hand, realizes that for the production of these alcohols, hardly any expertise is needed. Simply add your koji to a grain and let time do the rest. Everything else is flexible; once you have added the koji to a mixture of grain or cereal and a bit of water, there is very little you can do to stop it from turning into some kind of alcoholic beverage, and by extension vinegar. This is a natural process so old that it predates our own evolution. Vinegar just naturally happens.

What follows is a discussion of some of the different styles of alcoholic beverages that can be made using various kojis. From the widely known sake of Japan to the more obscure *tapai* of Malaysia and Indonesia, we'll provide a guide to making these different beverages. We'll also dive into the simplicity of making vinegar from these beverages.

All koji-based fermentation was developed in Eastern Asia, specifically China, and spread throughout the surrounding areas. John Hutt points out that due to various histo-socio-political reasons, those of us in the English-speaking world, especially in the United States, are more familiar with the works of Japan and sake than we are with those of China or Korea, the other large areas in the region, let alone the entirety of Southeast Asia. Physical borders such as mountain ranges and vast expanses of water are very real dividers of cultures, histories, and various koji spores, while political borders are porous by their willed-in existence. Does the People's Republic of China use a notably different set of fermentation techniques than Macau or Taiwan? No. Extrapolate that: Does South Korea? Does North Korea? Japan? We are all united in using techniques based on the availability of products, designing fermentation vessels based on our given climates, and our individual yet seemingly unifying religious adherence and associated dogma. This is one of the brilliant beauties of these beverages, and their differences fade after their labels are worn away.

## Choosing Your Starter

Starters for alcoholic beverages are like sourdough starters in some regards. They are hung, dried, inoculated with the environment, then stored for

## SAKE, SHOCHU, AND AWAMORI

Stephen Lyman

At some point in Japan during the Asuka period (530–710 CE) or the subsequent Nara period (710–794 CE), it appears that Buddhist monks managed to isolate koji, yeast, and lactic acid bacteria to develop modern sake production methods. While sake production continued to evolve over the intervening thirteen hundred years, the foundations of the process were in place by the 800s and sake good enough to be considered a commodity was being sold by the early 900s. How and why the Japanese monks decided to isolate the organism will remain a mystery, as no written records of the development process remain.

Once rice is polished, inoculating a batch of steamed rice is the first step in the process of making sake (save for the extremely traditional and rare *bodaimoto* style, which adds koji later). This inoculation process takes about forty-eight hours and ends when the koji rice is moved to either a starter mash (in the *kimoto* or *yamahai* processes) or directly into the fermentation tank (in the most efficient *sokujo-moto* process). The purpose of the bodaimoto, kimoto, and yamahai processes is to cultivate natural lactic acid bacteria in order to protect the later fermentation from corruption by other organisms. The koji used for virtually all sake production, yellow koji (a variant of *Aspergillus oryzae*), does not create much if any acid on its own, and so sake ferments are particularly sensitive to infiltration. This is also why most sake is fermented at very low temperatures where other organisms have a harder time taking hold. In the modern sokujo-moto process, commercial lactic acid is added to the fermentation from the beginning, which saves weeks of production time for each batch. It also results in a more consistent product since the brewmaster is relying on the work of only two living organisms rather than three. The upside of the low-acid yellow koji sake production process is that the resulting beverage is clean and sweet, and can be enjoyed right after pressing and filtration to remove residual rice particles.

A full seven hundred to eight hundred years after modern sake production methods were established, distillation technology arrived in Japan. It's unclear if this came from trade with Korea, China, or Okinawa (the independent Ryukyu Kingdom at the time), but the pot still quickly changed the drinking habits of Japanese living in warmer parts of the country. Warm-climate sake ferments that may have been corrupted with other organisms could be distilled to make something more palatable. It's believed the very first shochu was distilled sake in Kyushu in southern Japan. This was very much farmer's moonshine, the production of which was not strictly regulated by the government until the late 1880s when shochu producers were forced to incorporate. As a result, most of the oldest shochu distilleries date from 1884 or 1885.

That's not to say it was a free-for-all. Japanese peasants in Nagasaki Prefecture switched to making barley shochu because the local samurai overseers punished them for making alcohol out of rice, which was how taxes were paid during the Edo period. Likewise, farmers in the Satsuma Domain (modern Kagoshima) began making shochu out of sweet potatoes shortly after the spuds arrived from Okinawa via China via South America thanks to Portuguese traders.

But shochu production was by no means limited to southern Japan. Sake lees (rice solids left over after sake production) can be used as fertilizer if the remaining alcohol is removed. This was done by distilling the lees directly. At one point nearly every sake maker in Japan was also making *kasutori* (sake lees) shochu and making it available to the local townspeople as a cheaper alternative to their expensive sake, which by the Edo period (1603–1868 CE) was often being shipped off to Edo (Tokyo). They'd then further offer the now alcohol-free lees to local farmers to increase rice yields for next year's rice plantings. A very nice circle of life, which maximized use of an agricultural product on the island nation.

At some point—and again, nobody knows exactly how or why—black koji (*Aspergillus luchuensis* var. *awamori*) arrived in Kyushu via Okinawa, where it had been used in alcohol production for centuries. The name comes from the mold turning dark gray once the koji goes to spore. This never took off in sake production because black koji creates a lot of acid and also imparts strong earthy flavors and aromas that would not be appropriate in the elegant sake. However, once it goes through a still, it tastes just fine.

In 1923 a further evolution of koji occurred when Professor Genichiro Kawachi discovered a mutation of black koji growing in his laboratory. This mutation, *Aspergillus kawachi*, turned bright white rather than black. When used to make alcohol, it still created a lot of acid to protect the fermentation, but did not impart the earthy aromas associated with black koji. Today most shochu is produced using this white koji because it results in an easy-drinking beverage while the aromas of the base ingredient can shine through. It also has the added advantage of not growing all over everything in the breweries.

*Okinawan awamori* (a type of distilled spirit) predates shochu by at least one hundred to two hundred years, and its distilling tradition can very clearly be traced back to Ryukyu Kingdom trade with Siam (modern-day Thailand) where *lao khao* was the drink of choice. Originally, awamori was reserved for the Okinawan king and his court. In fact, illicit production of the spirit was punishable by death. Due to this tight regulation, only one district in the capital city was licensed to make awamori—and it was within sight of the castle walls.

All awamori was traditionally made with black koji, *Aspergillus luchuensis* var. *awamori*, (Japanese yellow koji, *A. oryzae*, may have been unknown in the kingdom at that time) and aged in clay pots to mellow the raw spirit. It's believed to have been primarily made with Thai rice, although there are records of other fermentable grains or tubers being used. As for the clay pots, the Okinawans developed a Spanish sherry-like solera method. In this process, if a customer ordered awamori, their personal vessel was filled with spirit from the oldest pot of awamori the producer had; then that pot was refilled with spirit from the second oldest pot and down the line until newly made awamori was put into the youngest pot to replace what had been sold. There are records of 200- to 250-year-old awamori existing into the early 1900s, but unfortunately all of those stocks were lost during the Battle of Okinawa during World War II. Remember, all of the producers were within sight of the castle walls—a castle that had been commandeered by the Japanese military for defense during the invasion. In fact, it was believed for several years that black koji had also become extinct as a result of the extensive American bombing, but then some of the mold was found growing on a few grains of rice found on a straw mat buried in the rubble. Awamori production was only resumed after the occupying American forces allowed Okinawans to resume using rice for alcohol production after 1948.

Today all awamori is made with long-grain Thai rice imported from Thailand and can be made only with black koji. As a result, awamori has geographic indication protection from the World Trade Organization. And thanks to that black koji, the walls and ceilings and every spare nook and cranny of awamori distilleries are covered by the ancient mold.

Most simply, koji is the Asian alternative to malting in the process of making alcohol from starches. However, koji adds flavors and aromas that are absent from Western traditions. It's a fascinating world to explore, and just when you think you understand it, you learn something that turns your understanding on its head. For example, some sake is now being made with black koji, yet the acidity is virtually absent. There's a shochu maker in Kagoshima that's now using yellow, white, and black koji in the same fermentation. And an Italian craft beer maker produced a yellow koji beer that tastes almost like a mead, it is so sweet and honeyed. Yet the same brewer's black koji beer tastes like the world's most potent sour. All in all a fascinating world to explore through the bottom of a glass.

later use. Some are easier to create than others. All are unique. The various names koji alcohol starters have in Asia differ as much as the individual cultures and languages do. While they are all one species, they can essentially be regarded as different variants that have been optimized to perform specific tasks via selective breeding by the individuals who lust after them. John Hutt has provided us with this breakdown:

*Qu*—China
*Koji*—Japan
*Nuruk*—Korea
*Murcha / marcha*—India and Nepal
*Banh men*—Vietnam
*Paeng*—Laos
*Loog pang*—Thailand
*Mochi kouji*—Myanmar
*Mae domba*—Cambodia
*Ragi Manis*—Indonesia and Malaysia
*Bubod*—Philippines

## How to Make Alcohol

To make an alcoholic beverage, you need little more than sugar, water, and yeast. In many instances, such as wine making, these three things can all be found within and on the grape itself. All the winemaker has to do is crush the ripe fruits and let the mash hang out for some time while the yeast metabolizes the sugars in the juice into alcohol. It really is that simple. You could easily do the same thing with a barrel of sugar water. You don't even need to add the yeast—it'll find its way into the sweetened water eventually.

When making the alcohols traditionally made with koji, simply pitch an amazake with yeast, and you will eventually end up with an alcoholic beverage. Realize, however, that you may not end up with a *good* alcoholic beverage. Though what one person deems to be divine another can call disastrous, we can safely state that the production of good, quality, enjoyable alcoholic beverages requires a fair amount of skill. This also holds true for the production of vinegar. If you start with a disgusting alcoholic beverage, chances are that you will end up with a disgusting vinegar. You can freely test this out by making a vinegar out of cheap wine or beer, but you won't want to do it again.

As we previously discussed, sake makers and the makers of other koji-based alcoholic beverages must take intimate ownership of every available optimization and variable if they wish to create a delicious and enjoyable beverage. From balancing the production of various enzymes during the cultivation of the koji to ensuring that the mash and lees don't become contaminated with a species of *Lactobacillus*, even the slightest imbalance can spell disaster. Off-flavors and -aromas coupled with unbalanced sour notes can easily be picked up on by even the most uninitiated of palates when sipping on one of these beverages, as they are often enjoyed and judged on their own merit as a singularity. Any and all changes, from ingredients to temperatures to geographic location, will result in a different product. Hence you have the dozen, or three dozen depending on whom you ask, types of sake based on when you inoculate the rice, when you filter the rice, what type of rice you use, and so on, with endless variations. You can multiply this diversity by a hundred when we are talking about preparation of Chinese *huangjiu*, or yellow alcohol, because it can be made of any grain, or indeed any fruit.

While not always the case, in order to easily brew an alcoholic beverage from koji, you'll need amazake first (follow the steps outlined in "Amazake" on page 112). Once you have amazake on hand, the following steps will guide you to making the simplest alcoholic beverage. We will discuss various optimizations and paths to brewing as the chapter progresses, followed by how to turn your beverage into vinegar.

With the following process, you can make as little or as much alcohol as you want:

1. Fill a widemouthed bucket nearly to the top with amazake.
2. Cover it with cheesecloth.
3. Allow the bucket to sit at ambient temperature until it becomes alcoholic.

If you were expecting more methodology, then you are out of luck. This process truly is that simple! That doesn't mean that you can't tailor and optimize your process to suit your own wants and needs. Again, it also doesn't mean that your alcohol will be delicious. You may not find the wild strain of yeast that cultures the amazake to be delicious. You could also end up culturing *Lactobacillus* in the amazake, resulting in sour amazake. Here are some exploitable variables that you can tamper with to create a nearly infinite array of possible outcomes:

- Change the substrate you grow your koji on.
- Use different variants and species of koji.
- Fortify the brewing alcohol with various herbs, spices, extra sugars such as honey, and fruits or vegetables.
- Use different water-based liquids to make the amazake mash. For example, you could use apple juice and end up with a hard sake cider.
- Experiment with holding the brewing alcohol at different temperatures.
- Use different strains and species of yeasts instead of relying on ones captured from the wilds of your larder. Swing by your local brewing supply store and see what they have available. Use a carboy with an air lock when doing this so that your selected yeasts aren't contaminated with unwanted ones.
- Introduce a lactic acid starter to your brew toward the end of fermentation. Lacto-fermented ginger is great for this.
- Condition your bottled alcohol to carbonate it or age it.

How do you know if your alcoholic beverage is ready? Once again simplicity reigns. Taste it. If you like it and enjoy the amount of alcohol, then get it cold and bottle it to arrest further fermentation both by the yeasts and by species of *Acetobacter*. If you are curious what the ABV is, then you can buy an instrument from a brewing supply shop called a hydrometer. This instrument will allow you to accurately measure the ABV. Follow the instructions that come with the device. If you just want a ballpark estimate, then crack open a beer with a 6 percent ABV and drink a sip of your brew against the beer. If the alcoholic "burn" feels the same, then you will have an idea as to where your ABV is. If it feels more intense or less intense, you will still have an idea as to what the ABV could be.

Chef John Hutt has brewed many of the traditional styles of alcoholic beverages made with koji found throughout Asia, especially China, the birthplace of *Aspergillus oryzae*. He states that the perception of impenetrability and years of tradition can stand in the way of what is ultimately an easy and user-friendly technique: the making of alcohol. These methods and techniques can be refined in myriad ways and will adapt to your surroundings naturally. John recommends that you approach the creation of drinks with the understanding that you are never going to make Fujian Laoji, Moutai, or some hyper-stylized sake unless you go make them in those places with those tools and specific microbial terroir. You should still realize, though, that what you make will be eminently drinkable, cookable, and yours—so enjoy it to its fullest. We will now offer a concise breakdown of established styles and the methodology behind them. First is John's method that he uses in New York, and then we'll hop across Asia.

# John's Rice Wine

John's wine recipe is essentially a version of *makgeolli* that he pushes to a higher ABV. In *The Art of Fermentation*, Sandor Katz recommends throwing a sweet potato into makgeolli. John uses purple Japanese yams, which makes this alcohol look like a grape-based red wine. You'll love the simplicity of making this and will undoubtly want to always have it on hand.

**4 L (1 gallon) cooked rice**
**2 cooked purple yams, large**
**1 L (1 quart) water**
**1 qu (Chinese yeast) ball or other *Aspergillus oryzae* starter such as nuruk**

Allow all the ingredients to cool to approximately 100°F (37°C). Mash and mix the ingredients together in a plastic bucket. Cover with cheesecloth and allow this mixture to sit for 3 to 4 days at room temperature. Mix and strain out solids using a brewer's bag filter (this can easily and inexpensively be found at your local brewing supply store or online). Place the liquid in an 8-liter (2-gallon) clay pot with an airtight lid for aging. A large glass jar or plastic food-safe bucket of the same size will also work should you not have a clay vessel. If air is allowed to get into the vessel, then it can, and will, further ferment into vinegar. Age for up to a year and then enjoy or bottle for later.

# Korean Makgeolli

Makgeolli is what happens when nuruk (see the "Meju" sidebar on page 141) is mixed with cooked rice or starches and allowed to liquefy into a porridge-like consistency. The steps are pretty basic, as is the way makgeolli is enjoyed: You can easily find street vendors in Seoul hawking it.

**500 g (1 pound) cooked rice**
**500 g (1 pound) water**
**500 g (1 pound) nuruk**
**1 sweet potato, large (optional)**

Cook the rice until al dente. Allow it to cool to about 100°F (37°C), then add your nuruk and mix well. Grate and add the sweet potato if you're using one, and again mix well. Let the mixture sit in a warm place for 2 to 3 days. When served unfiltered it is makgeolli. Should you want to refine it, then filter out the top layer of clear wine to make *cheongju*, or distill to make *soju*.

# Indonesian Tapai

Tapai (also called *tape*) is an alcoholic, very thick porridge-like food that has been cultured and fermented with the Indonesian koji starter Ragi Manis. It can be enjoyed by as is, garnished with various pickled or grilled fish and shellfish, or can be filtered and drunk as a delicious boozy beverage. This beverage is made from various starches, depending on where you are in Indonesia. You can select one of the starches mentioned below or use a combination.

**500 g (1 pound) rice, sweet potato, or cassava**
**25 g (0.88 ounce) Ragi Manis**
**50 g (1.76 ounce) palm sugar**

Cook the starch of your choice. The rice should be al dente and the sweet potato or cassava should be easily pierced with a fork. Allow the starch to cool to about 100°F (37°C). If you're using sweet potatoes or cassava, cut them up into small pieces about the size of a marble. Sprinkle the Ragi Manis on in an even laver over the starch and mix gently to coat all of it. Evenly layer the inoculated starch into a non-reactive vessel such as a mason jar, alternating between starch and sugar until the vessel is full. Ferment in the covered container for 2 to 3 days in a warm place. Once it's fermented you can eat it or mix with an equal part water to make a drink. The result is delicious and mildly alcoholic.

## JIUQU RECIPES

It is important to note that as of this writing, alcohol production in China is not standardized, so there are multitudes of types and terms for various processes. Several types of starters for cereal wines are all gathered under the heading jiuqu. Xiaoqu and Daqu, and their variations, are both considered to be Jiuqu.

# Daqu

A substrate, normally wheat, but sometimes barley and pea flour, is inoculated with *Aspergillus oryzae* and various species of yeast, then formed into blocks and dried. These starters are used to make a plethora of different styles of brewed and distilled alcoholic beverages. Each style is unique and directly reflective of a specific geographic region in China. There are many different types of Daqu, all of which can be cultured using the specific species listed below. (To make a Daqu fermentation starter, just swap out *Rhizopus oryzae* for the specific mold and follow the instructions in the Xiaoqu recipe, page 176.) We highly recommend that you experiment with all the different types of Daqu and find the one or ones you enjoy the most.

**Red Daqu:** Inoculated with *Monascus purpureus*
**Yellow or Green Daqu:** Inoculated with *Aspergillus oryzae*
**Gray or White Daqu:** Inoculated with *Rhizopus oryzae* or *R. chinensis*
**Black Daqu:** Inoculated with *Aspergillus niger*

## Xiaoqu

Often found at Asian supermarkets and called Shanghai Yeast Balls, Xiaoqu are 100-gram (3½-ounce) rice dough balls allowed to dry for a few days after being inoculated with *Rhizopus oryzae*. You can easily purchase them or follow these steps to make your own. This methodology calls for using rice hay. Seeing that obtaining rice hay is nearly impossible for those of us who live outside rice growing regions, you can easily source the spores of *R. oryzae* at most Chinese markets to use as a culture. The finished balls are then mixed with cooked rice to ferment into alcohol. We typically use three balls per kilo of cooked rice when making this style of alcohol.

**500 g (1 pound) rice flour**
**500 g (1 pound) water**
**1 kg (2 pounds) rice hay or 10 g (0.35 ounce) *Rhizopus oryzae* spores (dispersed)**

Make a rice dough by mixing the rice flour and water. If you're using spores, add them to this mixture. Roll the rice dough into balls that each weigh about 100 grams (3.5 ounces), about ten. Allow the outside of the balls to dry slightly and then place them in an incubation chamber (a box with some rice hay if you're using hay or a setup of your choosing detailed for growing koji) for 4 to 5 days at around 77°F (25°C). After this time the balls will be fully cultured. Remove them from the incubation chamber and let them slowly dry at no more than 100°F (37°C), after which they will last indefinitely. To use the balls to make alcohol, simply crush them up and mix them with cooked rice. Place this mixture into a glass jar and allow it to ferment for up to a week. After this you can strain and filter the solids out and then drink the alcohol, use it to cook with, or further ferment it into vinegar.

## Vinegar

Vinegar is a transformative ingredient. It has the ability to not only preserve our food but also easily elevate a bland or mundane dish into something blissfully bright and transcendent. For Larder and other restaurants that source local food and are far removed from the areas that produce citrus, vinegars have become the go-to acidifier for virtually any gastronomic application. One of the truly great things about vinegar is that it is just as easy to customize it and specify its use as it is to create a "cure-all" style that can work with any set of ingredients.

Vinegar is as ancient as alcohol, and like alcohol we will never be able to truly pinpoint its origins or birthdate.[1] It's safe to say that for as long as alcohol has been made and enjoyed, so has vinegar. Vinegar also holds many of the same religious, cultural, and economic significances that alcohol does; simply read a religious text of your choosing and chances are that it will mention vinegar. People the world over have used it for everything from medicine to a food preservative. The link that binds alcohol and vinegar is indisputable. Simply: Vinegar is made from alcohol. This process is so natural and virtually automatic that for centuries inventors have been tinkering with ways to *prevent* alcoholic beverages from fermenting into vinegar. From the narrow-necked amphorae of ancient Greece and Rome to newfangled contraptions that inject argon gas into your bottle, however, nothing has been able to completely stop this. As soon as alcoholic beverages with low alcohol percentages (6 to 12 percent ABV—lower will lead to weaker vinegar) are exposed to air, the widespread and ubiquitous bacteria *Acetobacter* will colonize the liquid, and through fermentation convert the alcohol into acetic acid, vinegar.

Vinegar's uses in the kitchen should never be understated. From pickles to sauces and desserts to cocktails, you can find a place for vinegar in any cuisine. Jeremy has been sober for a number of years and thoroughly enjoys being able to sip a vinegar-based mocktail that still recalls its origins. Vinegar will allow you to not only preserve your food via pickling but will also allow you to heighten the flavors of your food. Acids act to brighten, bolden, and harmonize the taste of a dish; many chefs add vinegar to a stock or sauce when "something's missing."

One of the joys of making and using vinegars made either with koji or from alcoholic beverages that were made with koji is the depth they contain compared with other styles. The vinegars of China, Japan, and Eastern Asia in general are known for having both a thicker viscosity and a more fantastic underlay of umami than those made from European-style alcoholic beverages. This is in part due to the breakdown of proteins in the substrate that the koji has grown on into tasty amino acids. This extra layer of complexity allows for what we feel is a broader use of these vinegars in your kitchen. They'll pair just as well with savory dishes as they will sweet ones.

## MAKING VINEGAR

There really is no culinary task that you could tackle that is simpler than making vinegar. It's even easier than brewing alcohol and will eventually happen on its own if an alcohol of the right ABV is left exposed to oxygen. Species of *Acetobacter*, most prominently *A. aceti*, live and thrive in alcohol. They're not just able to tolerate it but use it as their main source of food. These bacteria metabolize alcohol into acetic acid; this acid, in combination with the residual water and other components of the alcohol, is what we call vinegar. To obtain a culture of *Acetobacter*, all you really have to do is let alcohol sit out exposed to oxygen. Eventually the bacteria will colonize and form a raftlike matrix made up of cellulose. This raft looks and feels like a firm jelly and is referred to as a mother of vinegar. Should you want to speed up this process, you can easily source and purchase a mother by buying it online or even from your local brewing supply store. You can also source raw unpasteurized vinegar, which won't contain an apparent mother but will have plenty of bacteria in it to grow into one, the process of which we detail next.

In general, all you need to make vinegar is an alcoholic beverage with an ABV of 6 to 12 percent, a widemouthed vessel, and enough cheesecloth to cover the mouth of the vessel. Simply pour the alcohol into the vessel, cover it with the cheesecloth, and wait. Depending on the alcohol, the

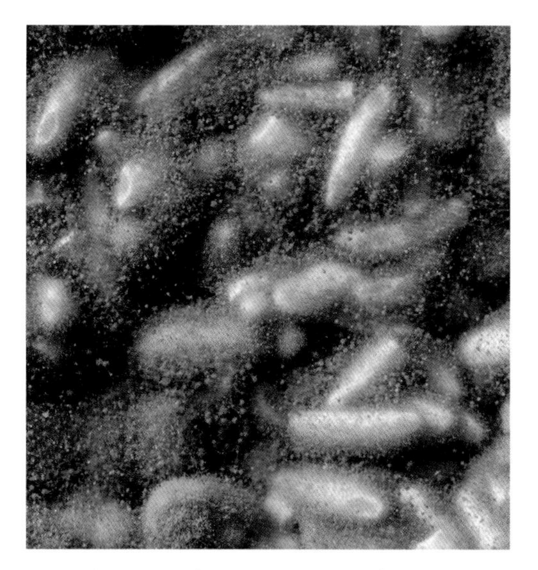

Spored-out koji (*Aspergillus oryzae*) at the end of its life cycle. *Photo by Peter Larson.*

Landscape of spored-out koji (*Aspergillus oryzae*). *Photo by Peter Larson.*

An array of different types of koji spores. *Photo by Peter Larson.*

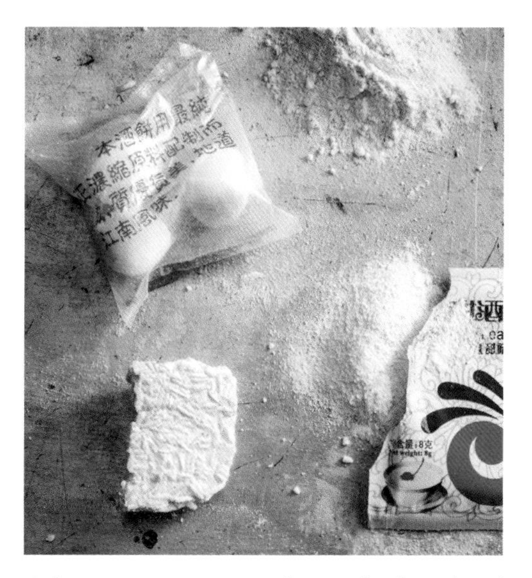


Other common starters: *Rhizopus* (*right two*) and jiuqu (Chinese yeast balls, *top left*) with jasmine rice koji (*bottom left*). *Photo by Peter Larson.*

A dehydrator with a multitude of koji makes in progress. *Photo by Peter Larson.*

Popcorn koji. *Photo by Andrew Wang.*

Spores on popcorn koji. *Photo by Andrew Wang.*

Water bath incubator setup. *Photo by Andrew Wang.*

Jasmine rice koji at 0 hours. *Photo by Peter Larson.*

Jasmine rice koji at 12 hours. *Photo by Peter Larson.*

Jasmine rice koji at 24 hours. *Photo by Peter Larson.*

Jasmine rice koji at 36 hours. *Photo by Peter Larson.*

Caramelized jasmine rice koji. *Photo by Andrew Wang.*

South River Miso's dried brown rice koji. *Photo by Andrew Wang.*

Oaxacan blue corn amazake. *Photo by Andrew Wang.*

Nixtamalized corn koji using Castle Valley Mill's bloody butcher corn. *Photo by Andrew Wang.*

Sweetfish roe amino paste. *Photo by Andrew Wang.*

Husk Savannah's amino pastes: tomato, chanterelle, and Hubbard squash. *Photo by Andrew Wang.*

Koji for an amino sauce of black soybeans and Castle Valley Mill's spelt. *Photo by Andrew Wang.*

An assortment of amino pastes in jars. *Photo by Peter Larson.*

Tamari at the top of a matzo ball amino paste. *Photo by Peter Larson.*

An assortment of amino sauces. *Photo by Peter Larson.*

Spent beer grain amino sauce. *Photo by Peter Larson.*

Amino paste made from Sub Edge Farm's sunflower husks, a by-product of making sunflower seed oil. *Photo by Andrew Wang.*

Umami chili oil using sunflower barley douchi as a base. *Photo by Andrew Wang.*

Kim Wejendorp's carrot scrap mirin. *Photo by Andrew Wang.*

Koji-cultured beet charcuterie. *Photo by Peter Larson.*

Charcuterie: koji-cultured beef rib eye, amazake coppa, and amazake mettwurst. *Photo by Peter Larson.*

Applying koji starter to a raw steak prior to culturing. *Photo by Peter Larson.*

Koji-cultured steak before and after cooking. *Photo by Peter Larson.*

An assortment of koji fish charcuterie. *Photo by Peter Larson.*

Larder's vegetable charcuterie: beet, carrot, and daikon (*top to bottom*). *Photo by Peter Larson.*

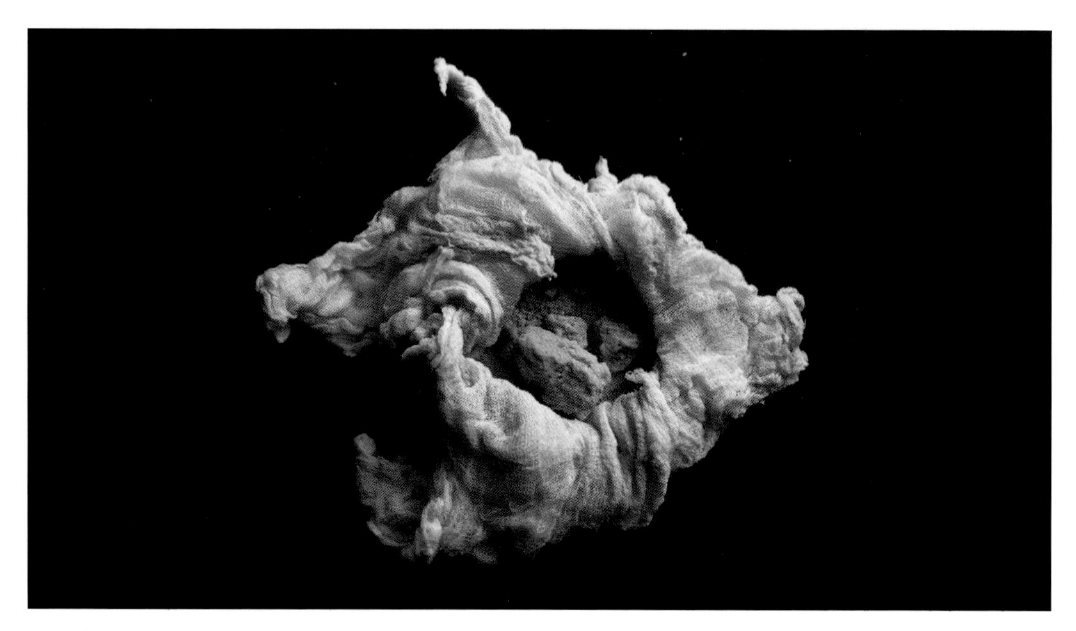

Dried ricotta miso cheese. *Photo by Claudia Mak.*

Yellow House's shio koji washed cheese. *Photo by Peter Larson.*

Kojizuke: sunchoke, golden beet, and bok choy. *Photo by Andrew Wang.*

Cultured Pickle Shop's golden beet kasuzuke. *Photo by Andrew Wang.*

Cherry peppers pickled in shio koji made with Sean Doherty's lacto-fermented maple hot sauce. *Photo by Andrew Wang.*

Squash dried hoshigaki-style, made into a misozuke using an amino paste of 898 squash. *Photo by Andrew Wang.*

Amazake rye and amazake buttermilk breads. *Photo by Peter Larson.*

An apple slice vacuum-packed with jasmine rice koji. *Photo by Andrew Wang.*

ambient environment, and how much *Acetobacter* is present, this can take anywhere from a few weeks to a few months.

This process can be sped up if you forcefully aerate your vessel, which can easily and inexpensively be done with an aquarium pump. Source a pump that can handle the volume of the vessel that you are using. The pump will force air, rich in oxygen, into the fermenting alcohol and work to speed up the rate with which the acetobacter converts the alcohol into vinegar.

Harry Rosenblum, a vinegar enthusiast and aficionado, wrote a wonderful book on the merits of vinegar and its uses called *Vinegar Revival*. For many years at his culinary emporium, the Brooklyn Kitchen, Harry has been teaching classes about how to make vinegar. Harry is also a lover of all things Japanese, and along with fellow vinegar obsessive Michael Harlan Turkell—who has also written an unbelievably fantastic book about vinegar called *Acid Trip*—hosts a wonderful event series across the country called Sumo Stew. At this event, Japanese food and drink are consumed while people are immersed in an evening of Japanese culture and watch live-streamed sumo wrestling. Harry feels that the best way to make a koji-based vinegar is to start with amazake. He simply adds in some yeast and then lets it ferment for a few weeks. By adding yeast (champagne yeast will do, as will wild yeast, or better yet more specialized sake yeast) and putting this in a sanitized jar or jug with an air lock, your amazake will turn into a simple sake within about a month. You will now have a rice-based alcohol that is perfect for converting to vinegar. Simply let the alcohol sit for longer, exposed to air, and it will eventually become vinegar. Harry's method for making vinegar is a fairly straightforward and common method, and you'll find a similar methodology being implemented across the world.

According to Harry, once you have vinegar, things can get sour in the best possible way. It takes the work of two fungi, one bacteria, and a bit of coddling and harnessing on our part, but the effort is worth it. You will have delicious vinegar that is so acidic, you can use it to preserve everything from meats to fruits and vegetables. Always remember to strain your vinegar, place it into a narrow-necked bottle, and start a new batch so you won't run out when you finish the first one.

## AWAMORI VINEGAR

### Alfred Francese

*Alfred Francese is the larder master at the award winning Emmer & Rye restaurant in Austin, Texas, where he works closely with chef and owner Kevin Fink to create myriad ingredients that are used on a daily basis in the Emmer & Rye kitchen. One of the fantastic things about what happens at this restaurant is its exclusive use of foods and ingredients sourced solely from the state of Texas. From snails foraged from the banks of the Colorado River to amazing produce sourced from HausBar Urban Farm in the heart of the city, Emmer & Rye is a prime example of regional influences on koji and its uses. This recipe is a stunner and will surely intrigue anybody that you serve it to. One of the things we love about it is its complex sour profile—not only from the acetic acid but also from the citric acid created by the* Aspergillus luchuensis *var.* awamori *used in its production. It's a double punch of sour power.*

*Aspergillus luchuensis* var. *awamori* creates a really unique koji for two reasons: one, because it creates a black web of mold to bind the rice together, and two, because it also produces citric acid. Traditionally awamori is a distilled spirit created from non-roasted black koji. However, once the koji is roasted, it has a warm aroma of cereal grains, a background of umami, and fruit notes of pear and apples from the citric acid. Transforming this into an alcohol creates a unique flavor where all of this is expressed. Preserving it as a vinegar means it can be enjoyed in many ways.

**4 kg (8.8 pounds) black koji grown on Carolina Gold rice**
**8 L (8.5 quarts) water**
**1 g winemaker's yeast**
**400 g (14.1 ounces) unpasteurized vinegar**

Roast the black koji at 350°F (176°C) for 20 minutes or until it takes on a roasted aroma and starts to caramelize. Wait until the koji cools and wrap it with cheesecloth into a large sachet. Bring the water to a boil, turn the heat off, and add the sachet. Cover the pot and allow it to stand for 1 hour. Once the hour is up, lift the sachet out and dip it back and forth into the water ten times. Use tongs or heat-resistant gloves to squeeze out any excess liquid. This is your wort.

Allow the wort to reach room temperature, and read a sample of liquid using a spectrometer. The liquid should be 9 to 10° Brix. Be mindful that rice will differ from farmer to farmer, season to season, and region to region. Differences may occur and can translate to different levels of sugar in the wort. If the sugar levels are not the same as in this recipe, you can easily increase your Brix by using raw cane sugar. Once the Brix is at the correct level, bloom your winemaker's yeast in a cup of lukewarm water for 15 minutes to activate it then pitch it into the wort and give it a light stir. Pour the wort into an air-locked carboy and allow it to ferment for 7 to 10 days at 80°F (27°C). Once the bubbling has slowed down, check the Brix again. It should be 0 to 1° Brix, and taste dry and alcoholic. This is the end of the primary ferment.

To obtain vinegar you'll need to ferment this alcohol. At Emmer & Rye we keep a raw white vinegar on hand made from distilled alcohol, but any neutral-flavored unpasteurized vinegar will do. Remove the lid from the carboy and back-slop the unpasteurized vinegar in. Check that the pH level is 3.5 to 3.8 using a pH meter or test strips. If it's not, adjust the mixture further with more raw vinegar. Transfer the vinegar to a large widemouthed vessel and cover with a secured linen. Check the pH of the vinegar every week until it has reached 2.8.

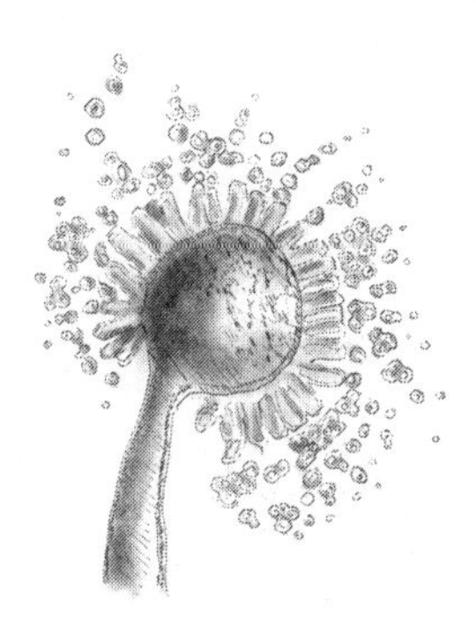

## CHAPTER 10

# Aging Meat and Charcuterie

Jeremy is one of those dreamers who isn't afraid to try something theoretically sound, but completely off the wall, to see what happens. This is how he came up with the idea of growing koji directly onto meat to make it more delicious. Using koji to age meat is a non-traditional method, and we are here to show you something different that is delicious and can easily be made.

## Aging Meat Primer

The hanging, or aging, of meat is an ancient practice, something that peoples in temperate climates have been doing for centuries. Early on humans realized that when an animal is killed for food, many changes to the meat occur both at the time of death and during the time between death and consumption, and so they developed the practice of hanging meat and

## KOJI AND MEAT: MAGICAL SORCERY

**Meredith Leigh**

*Meredith Leigh has worked as a farmer, butcher, chef, teacher, nonprofit executive director, and writer, all in pursuit of good food. She is the author of* The Ethical Meat Handbook *and* Pure Charcuterie. *She uses koji in the creation of her charcuterie and loves the new possibilities that it affords her in creating unique, one-of-a-kind cured meats.*

The first time I inoculated a piece of meat with koji, it was a 5-pound (2.8 kg) pork loin cured with salt and Chinese five-spice seasoning. It was a trial for an Asian-fusion restaurant I was working with, and I was absolutely certain that piece of meat was going to rot. Looking back, it still feels like the most ludicrous thing I've ever done: I bought a Tupperware container at the dollar store and an aquarium thermometer from the pet store. Then I filled the container with water, inserted a thermometer, and plugged it in. I put in one semi-expensive cut of meat, and then waited for magic to happen.

When I opened the container after thirty-six hours, I found not a putrid mass of pork, but instead a regular pork loin smelling something like a wildflower meadow, covered with the loveliest map of fungus in whites and pale yellows. I marveled, rejoiced, but quickly found other reasons to worry. Would the intense character of this koji mask the flavors of five-spice seasoning? Of the pork itself? I carefully weighed the meat, labeled it, and hung it up in my charcuterie cabinet.

In about *two-thirds of the time* I normally hang a lomo, the loin finished aging. I took it out, cut into it, and found not only that it had cured successfully and retained the character of the pork itself, but both the five-spice flavor *and* the undeniable beauty of salted fat were merely amplified by the koji. The sweet, salty, umami, and other flavors simply kicked up several notches. Thrilled, I took it to a food show, where chefs ate sample after sample, trying to understand what sorcery of flavor this was. They loved it.

Using koji to tenderize typically tough cuts of meat and impart a dry-aged flavor without the time, cost, and potential waste of standard dry-aging procedures can revolutionize the way we use whole animals. Just as charcuterie is a way of using meat creatively and fully (and has been redefining the way we incorporate protein in the diet), koji offers the same avenues through its uses in fresh cooking as well as fermentation and curing.

Since discovering koji, I've jumped headlong into the many implications for charcuterie and meat cookery this amazing mold has to offer: for example, using koji's secondary ferments to create brines and probiotic inoculants, speeding up fermentation dynamically (without single-strain commercial starter cultures), and even attempting to harness koji's electromagnetic charge to create charcuterie using less salt. In my ongoing classes I am attempting to share this information liberally, and I urge students and readers to test it along with me. The community around koji continues to grow, and the synergy created by the people using it is not unlike koji's own enzymatic activation and change.

My experience with koji charcuterie continues to amaze me, and I have so much yet to learn. I will continue to work and also tap into the knowledge being built by the incredible cohort of chefs and home practitioners who are testing, creating, and open-sourcing uses for this amazing mold. I believe that our knowledge and use of koji is a scratch on the surface for what this organism can do for good food. In the quest for pure, healthful, and ethical meat and charcuterie, I know koji has complex and rich implications.

letting it age before consuming it. For them, hanging and aging meat was the way to go for three main reasons: flavor, texture, and necessity. These first two reasons still hold very true for those of us who have great access to refrigeration, and the third for those who don't. Aged meat is fantastically delicious and highly coveted by chefs, butchers, and diners alike.

Meat that is aged in traditional ways, however, is expensive and wasteful. First, due to the length of time needed to age the meat, it can lose 30 percent or more of its starting weight from water loss alone. This coupled with the amount of dried-up and molded meat that needs to be trimmed from the outer surface can amount to a product loss of around 50 percent—and that's before final-cut fabrication and cooking, which can result in additional loss. However, age-accelerated meats created using koji offer several benefits over traditional dry-aged meats, the first being that they have a nearly 100 percent yield. While the age-accelerated techniques we've developed using koji aren't the same as traditional dry aging and result in a different product, the changes in texture and flavor are similar enough that a comparison can be made when describing the final product, which can be just as funky, cheesy, and intensly delicious.

Many cultures have been using amino pastes and amino sauces for centuries to embolden and flavor their proteins. From steak to fish, there are plenty of opportunities to work koji into the foods that you create. Some of these techniques are as simple as rubbing an amino paste on a fish fillet and letting it marinate for some time, while others are slightly more complex and seek to optimize the activity of the enzymes koji produces to work their magic, such as on a steak.

We'll first explain some of the hows and whys of the traditional and industry-standard aging practices to give you a better picture of the reasons for aging meat and the level of commitment needed to age it. We'll also discuss charcuterie making from a traditional standpoint and how various expressions of koji can be used to create it. Following that we'll discuss how the use of koji and the techniques we've developed and adapted for use on animal protein can be put to a wide range of uses for you in your kitchen. And finally we'll look at using these same techniques to create the vegetable charcuteries Jeremy developed at Larder.

## THE HOWS AND WHYS OF AGING MEAT

Increased scientific comprehension of the world around us and the advent of storage technologies afforded us a greater understanding of the act of aging meat. We eventually understood the mechanics behind aging and its importance, and were able to comprehend the reasons aged meat tastes better and has a more desirable texture. And so, while aging meat started as a form of preservation, we also now understand how it impacts flavor and tenderness.

Many chemical and enzymatic changes occur when meat is aged. The most drastic change when going from live animal to meat, outside of slaughter and death, is a stage of death known as *rigor mortis* (Latin for "stiffness of death"). This happens to meat once all the available oxygen in a body system is depleted and an organic chemical that depends on oxygen to control muscle contraction and relaxation—adenosine triphosphate, or ATP—is degenerated. Due to this reaction the *myofilaments* (muscle fibers) stiffen and won't relax until enzymes, produced endogenously or coming from outside sources such as bacteria or fungi, work to break down the structural components of muscle and thus allow for a relaxation of the muscle tissue. During rigor mortis the ATP is used up completely, causing the myosin heads to remain locked to the actin filaments. ATP is needed to cause them to release (allowing the muscle to relax), but with the ATP depleted, the muscle becomes stuck and rigid in its last position. Meat that has been butchered, cooked, and eaten while in rigor mortis is tough and grainy, and can have a metallic ironlike flavor that virtually all meat eaters find off-putting. The good thing is that rigor mortis swiftly passes; after about three days, enzymatic action has begun to degrade the myofilaments that make up the muscles in the meat. This action breaks down the myofilaments while acids / alkaline proteases are activated and are responsible for most of the changes in the ultrastructure of the myofilaments. This results in meat that is more tender. It also results in carbohydrates, fats, and proteins being broken down into sugars, fatty acids and esters, and amino acids, respectively. These are what help to change the taste and flavor of the meat.

Years ago Jeremy had the opportunity to learn about this process firsthand. As a culinary student in the Hudson Valley, he was presented with the chance to participate in the slaughter and butchery of a cow. Since then he's personally slaughtered and butchered many animals, from pheasant and crow to venison and hog, and has seen the importance of properly aging meat. We firmly believe that aged meats are considerably better than unaged.

In today's foodscape, aging is done through a variety of processes for different lengths of time depending on the animal. The big three—chicken, pork, and beef—are generally aged for seven, twelve, and twenty-one days, respectively, before they arrive at a retail outlet. This is an industry-wide practice not only in the United States but in many other countries with centralized food systems. Some of the techniques applied to meat to age it include:[1]

**Pelvic Suspension.** When a carcass is suspended by its pelvic area and allowed to hang, gravity works to physically stretch and tenderize meat by pulling on it.

**Electrical Stimulation.** The meat is stimulated with high-voltage electricity to speed up the rate of glycolysis (onset of rigor), causing the ATP to be degraded more quickly; when the carcass is cooled, this protects the muscle from cold-shortening and becoming tough.

**Dry Aging.** The most common technique used by chefs and proper butchers. Meat is set in an environment in which airflow, temperature, and humidity are precisely controlled to allow the meat to age without spoiling. Many chefs and butchers agree that this is the aging technique that leads to the best and most enjoyable meats. The downside is that you lose a considerable amount of muscle due to dehydration and the effects of molds and bacteria, which have to be cut off and discarded.

**Wet Aging.** The technique that most meat packers and processors prefer. Meat is vacuum-sealed in a package and allowed to age in its own purge (the residual moisture that's left inside the myofilaments after all the blood has been drained). This is a very efficient way to age meat and does save money during the aging process because you can store more meat in a smaller space.

**Modified Atmosphere Packaging (MAP).** The meat is packed in a light brine, with the injection of carbon monoxide or nitrogen gases, or various combinations of these. This allows the meat to retain moisture, thus giving the final product a higher yield. It also gives bright red hues to the meat so that it appears fresh in the meat case at the supermarket.

Pelvic suspension and electrical stimulation are fairly standard and typically happen at the processing plant as rigor mortis is going on or right after it has passed. The second two practices are employed, or not employed, based either on the needs that a specific market dictates or on the preference of the chef, butcher, and their diners.

## DRY AGING MEAT AT HOME

You do not need a fancy setup or any major investment to dry age meat in the traditional way at home. It can easily and safely be done in your refrigerator. If you plan on dry aging meat in a restaurant or other business that sells food, you should check with your local regulatory agency to see if you need to have an HACCP (Hazard Analysis Critical Control Point) plan in place. Here are the things you need to dry age meat in your fridge:

- A sheet tray.
- Some sort of raised, non-reactive rack (stainless steel, for example, which won't rust) to rest the meat on.
- Large cuts of meat. If you decide to dry age a single steak, then you'll waste the whole thing. Due to the amount of water the muscle loses and the amount of meat you have to trim from the outside, you should only do this with subprimal cuts at a minimum.

Set the resting rack in the sheet tray and then place the meat on the rack. Put the tray in the area of your fridge that gets the most airflow and leave it be for as long as you plan to age it. Airflow is very important and will help keep the surface of the meat dry as it ages. Be sure to keep a record somewhere of what the meat is and the day you started to age it. This will ensure that you can properly manage the aging process.

Use clean hands to rotate the meat every few days so that a new side comes in contact with the resting rack. Follow the best practices for storing raw animal proteins in a refrigerator, as you don't want it to drip on your salad greens. Once the meat is aged to your liking, you need to trim and portion it. Trim it until you see brightly colored flesh and portion it to your liking. Because it's hard to know what kinds of molds or other microbes are growing on the surface of the meat, the trimmings are traditionally discarded. We have experimented at great lengths with these trimmings but do not advise you to do the same unless you are fully prepared to comprehend and deal with the consequences should some sort of pathogenic contamination have taken place.

Aged meat freezes well (see the sidebar "Koji Compared with Traditionally Aged Meats" on page 205); due to its low moisture content, it doesn't develop damaging ice crystals in its cells. This means that you can freeze it without worrying whether its texture will be negatively affected.

It should be noted that these techniques should generally *not* be used for poultry, especially chicken. While no process or environment for slaughtering and processing meat is perfect, the ones used for pork, lamb, and beef present less risk of contamination by pathogens than those used for chicken. This is due in part to the size of the animals and their entrails, which happen to be the main source of pathogenic contamination. Larger animals have larger and hardier entrails, which in turn are easier to safely handle during processing without breaking them. In an animal such as a chicken, the entrails can easily be disturbed and broken, and the fecal and digestive matter can come into contact with the meat. Once this happens the risk of contracting a foodborne illness increases dramatically. Chicken, and other intensively raised poultry, can also develop eggs that are contaminated with microbes, such as salmonella, before the egg is even laid.

## Curing Meat: A Primer

Cured meat, or *charcuterie*, has been made since the days of prehistory. Early versions of what we now eat for pleasure started as a way to preserve fragile muscular tissues that would rapidly spoil in the days before refrigeration. These prehistoric cured meats were primarily preserved via smoke drying. Long thin strips of meat would be butchered and hung over a smoky fire to dry and preserve them. The closest relatives we have to this style of curing meat nowadays would be the South African biltong and the various widespread styles of jerky.

As the Stone Age transitioned to the Bronze, marked improvements were made to the techniques used to preserve meats (and fish). Ingredients such as salt, saltpeter (potassium nitrate), vinegar, and various spices were added to meats not only to make them more delicious but primarily to make them safer and more stable over longer periods of time.

It is well documented that the inhabitants of Gaul (modern-day France) were curing whole-pig hams around the fifth century BCE, about twenty-five hundred years ago. These hams were heavily cured with saltpeter-rich salt and then smoke dried. Sue Shephard details this in her book, *Pickled, Potted, and Canned*:

> *The statesman Cato gives his own Rome-made ham recipe in his* De Agri Cultura, *the oldest surviving complete prose work in Latin. It directs the reader to lay a number of hams in huge earthenware jars covered with half a peck of Roman salt per ham. The hams are to be*

*left, with occasional turning, for twelve days, then cleaned off and hung in the fresh air for two further days. They are then rubbed with oil and hung and smoked for two days. Finally, they are rubbed all over with a mixture of oil and vinegar and hung in a meat store where "neither moths nor worm will attack it." Curing hams can be done with a range of mixtures based primarily on salt, with the addition of sugars and spices, herbs, and oils. Sugar and honey are also powerful preservers and have the added benefit of counteracting the hardening effects of saltpeter.*[2]

For societies around the world, the need to have shelf-stable nutrient-dense foods was second only to the need for clean water. If cared for properly, cured meat won't spoil or go bad, it's lightweight, it tastes great, and it is dense in essential nutrients needed to survive long periods away from fresh foods.

Two main areas of the world developed the mainstay styles of cured meats that we now enjoy today (this isn't to say that there aren't other areas of the world that developed styles of cured and preserved meats, only that these are ones now popularly enjoyed across the globe): the regions surrounding the Mediterranean Sea and those that make up what is now called China. Cured meats in both regions were buried in mixtures of salt and saltpeter (and spices) and closely monitored until they lost some of their moisture. They were then hung in slightly damp, cool locations or over a relatively cool, smoky fire to dry. With this process, depending on the cut of meat and who is making it, the level of dryness is equal to a weight loss of 25 to 50 percent of the starting weight of the meat before salting. It required an individual to invest large amounts of time into learning how long a given cut of meat should be buried and how often the salt should be washed off and reapplied. This technique is hard to master and control; in Parma, Italy, a person in charge of properly salting prosciutto is today named *Maestro Salatore* (salt master). Meats cured this way would frequently have to be soaked or boiled before being cooked; they weren't necessarily meant to be eaten on a charcuterie board. American-style country ham and Chinese salt-pressed duck are prime examples of this.

Eventually, sound scientific understanding of the curing process started to evolve and the equilibrium method (EQM) was devised and codified, which takes into account the rate at which salt travels through muscular tissue and the

## MORE ON CURING SALTS

Curing salts are various mixtures of salt, sodium nitrite ($NaNO_2$), sodium nitrate ($NaNO_3$), and food coloring (the food coloring is added solely to prevent confusing this mixture with seasoning salts). Pure $NaNO_3$ and $NaNO_2$ are lethally toxic in their pure forms, which is why it is nearly impossible to source them in purified forms without licensing from various government agencies. They do eventually break down into harmless compounds over the course of the crafting time. They retard spoilage and life-threatening foodborne illness such as botulism (this is most important in ground meat preparations). They are required by the USDA/FDA for various cured meats, especially ones made with ground meat, unless the product is labeled "uncured." They interact with residual myoglobin and oxygen to keep the characteristic red hues you see in cured meats. Cured meats made without nitrates and nitrites will rapidly oxidize and turn unappetizing colors of gray and brown.

Even "uncured" cured meats contain sodium nitrate and sodium nitrite. They don't contain the refined versions, however, but rather natural ones extracted from the celery plant. There is no difference in the chemical and molecular makeup of $NaNO_3$ and $NaNO_2$, no matter how the compound is obtained or synthesized. Therefore products labeled "uncured" are intentionally misleading and confusing. When used properly, curing salts will allow you to create not only beautiful cured meats but also safe cured meats that won't sicken or kill people.

There are two main formulations of curing salts, both of which can be easily sourced via a local purveyor, or online:

- Cure #1 contains 6.25 percent sodium nitrite and 93.75 percent table salt. It is recommended for meats that require short cures and will be cooked and eaten relatively quickly.
- Cure #2 contains 6.25 percent sodium nitrite, 4 percent sodium nitrate, and 89.75 percent table salt. The sodium nitrate found in Cure #2 gradually breaks down over time into sodium nitrite; by the time a dry cured sausage is ready to be eaten, no sodium nitrate should be left. For this reason it is recommended for meats that require long (weeks to months or even years) cures, like hard salami and country ham.

minimum percentage of salt needed to safely and efficiently cure meats. The percentage of salt and curing salts (see the sidebar) most often used are 3 percent and 0.25 percent, respectively. This means that if a 100 grams (3.5-ounce) piece of meat is to be cured it, it needs 3 grams (0.1 ounce) of salt and 0.25 gram (0.009 ounce) curing salts. Using the EQM will ensure that your cured meats will never be oversalted, and should never require soaking to desalinate them.

The equilibrium method is mainly used for whole-muscle cuts of meat but can also be used for ground styles. In the EQM, meats under 3 pounds (1.4 kg) and 3 inches (7.5 cm) in diameter are salted at the above-mentioned rates for a minimum of fourteen days. For pieces over 3 pounds and 3 inches (8 cm) in diameter, add two days for every additional pound of meat or every 0.25 inch (6 mm) in diameter. The great thing about the EQM is that due to using a set percentage of salt, you can essentially never overcure or oversalt your cured meat. Even if you forgot about the meat in the back of your fridge for a couple of months, it will never be too salty. Forgetting about a piece of meat can lead to other changes such as slight fermentation or pickling, but it will never be too salty.

During the salting process you can add spices and herbs to your cure. You may need to consult either existing literature or recipes to determine the potency and volatility of each spice or herb. Generally speaking a ratio of 4 percent of the weight of the meat is acceptable for most herbs and spices. Extremely potent ones such as chilies or juniper can be scaled back to 0.5 to 2 percent, while thyme or cumin can go as high as 6 percent. It really boils down to personal preference as to how much of a given spice or seasoning you like. Trial and error will guide you with the passage of time.

Once your meat is cured, it needs to hang in a place that will allow it to slowly lose moisture until it averages 30 to 40 percent weight loss. Use a piece of masking tape attached to a string or a charcuterie tag to label each piece of meat with the date that it was hung and its starting weight. Check it every one to two weeks on a scale and calculate its weight loss. Without investing extra money in a curing chamber, you can easily hang your meat in your refrigerator. Should you want to dial in your charcuterie to a more refined product, then either invest in a prefabricated curing chamber or make one with a refrigerator retrofitted with a humidifier. When you start to hang your charcuterie, you should have around 75 percent humidity and gradually decrease it to 50 percent as the meat hangs. Keep the temperature between 50 and 60°F (10–15°C), remembering that the higher the temperature gets, the greater the chances of spoilage or pathogenic contamination. One thing to note is that many people did, and still do, hang meat in their basement or cellar. In our experience you shouldn't take too much risk with hanging in an environment like this without fully comprehending what could go wrong without clean conditions and precise controls.

You will notice that your cured meat may develop a hard outer shell. This is called *case hardening* and can be remedied by either properly adjusting your humidity levels as the meat hangs or by vacuum-sealing the meat after it achieves the desired weight loss until the residual moisture redistributes from the center of the meat to the outer edge. Using vacuum bags to help equalize the moisture in your meat can take an additional few weeks. This doesn't always work, but most of the time it will help considerably. It also doesn't hurt to rub the surface of the meat with a flavorful alcohol such as red wine or whiskey before you vacuum-seal it to help hydrate the outer layer from the outside. Alcohols are best for this purpose compared with plain water or watery liquids such as juices or stocks due to their ability to inhibit pathogens.

From time to time you may notice that, if the humidity is too high in your curing chamber, mold can grow on your cured meat. While chances are that this mold is harmless, it is nearly impossible to distinguish illness-inducing strains from beneficial strains unless they have intentionally been inoculated. Should mold develop on the outside of meat that hasn't been purposefully inoculated, then make a paste out of vinegar and salt and use it to scrub the meat. This should only be done if there are no off- or offensive odors coming off the meat and only if the mold is just starting to appear. If large amounts of mold that you did not put on your meat are present, then exercise caution and your own judgment. You may want to consider starting over if it smells really off.

Should you want to *intentionally* culture mold on your meat (which aids in flavor development and texture, and acts as an additional biodefense against pathogens), then you'll need to source either the spores of *Aspergillus oryzae* (koji!) or *Penicillium nalgiovense* and inoculate the surface. (*P. nalgiovense* is marketed as Bactoferm Mold 600 and can be sourced via sausagemaker.com or Amazon.)

## LEVERAGING MICROBES IN CURED MEAT AND CHEESE

**Bentley Lim**

*Bentley Lim, PhD, is a researcher in microbiology at Yale University. Rich first met him when he was invited to teach at a fermentation intensive class at the Tsai Center for Innovative Thinking at Yale. Bentley was intrigued by what Rich had presented about the untapped potential of koji. As a result, he offered to support future work by providing his expertise in microbiology and the science behind fermented foods. Since then Bentley has played a key role in furthering our technical understanding to make better koji-based aminos. In this section we hear from him about the role koji, and other microbes, play as charcuterie is being made.*

The development of aromatic and umami-active compounds in miso and soy sauce also holds true during the fermentation of meat and cheeses.[3] For cheeses, lactic acid bacteria and other surface microorganisms provide the first wave of enzymes to digest proteins, specifically casein, into *peptides* (chains of amino acids) and supply nutrients to the second wave of microorganisms. Depending on how the cheese is initially processed, its ripening temperature, and the natural flora of microbes in the air, the next growth of microorganisms will determine the aroma, flavor, and texture of the final cheese product. In the process of curing meats, surface microorganisms supply a portion of the initial enzymes, but a majority comes from the enzymes found in the meat itself. As with cheese, the surface lactobacilli play a very important role, however; they produce the lactic acid that modifies the water-binding capacity of the meat proteins, thus contributing to texture, moisture content, flavor, and aroma of the final product. Following the drop in pH by the lactic acid bacteria, the secondary microorganisms, such as molds and yeasts, produce chemical compounds that contribute to much of the flavor and aroma typical of fermented meats.

In order to produce a consistent fermented cheese or meat product, companies have used a starter culture of microorganisms consisting of lactic acid bacteria, molds, yeast, and other microbes, resulting in a product uniform in texture, flavor, and safety. However, a key benefit of using a starter culture is the decrease in fermentation time required to achieve the final product. The requirements needed for the community of fermenting microorganisms to establish on meats or cheeses will depend on environmental conditions. By supplying a starter culture that would immediately start breaking down proteins, fats, and sugars, you can procure the taste of aged cheeses and meats in a much shorter time than when naturally done. This phenomenon occurs when treating fresh meats and dairy products with koji or shio koji, where *Aspergillus* provides the enzymes to break down proteins and fats into umami-active and aromatic compounds and to facilitate an environment for other microorganisms to produce additional flavor molecules. While the final flavor may be redolent of naturally aged meats and cheeses, much of the flavors and aromas come from the *Aspergillus* mold's metabolism of the amino acids.

However, with naturally aged products, taste compounds are generated with each wave of microorganisms. These taste compounds can be further altered with subsequent waves, thereby creating a very complex-tasting end product consisting of umami compounds, bitter compounds, and "kokumi"-active compounds that enhance continuity, mouthfulness, and thickness. Additionally, the multitude of different taste and aromatic compounds are likely to interact with one another and bring out more complex responses for the consumer.

## Curing Meat with Koji

One simple and straightforward way to incorporate koji into whole-muscle charcuterie and other cured meat preparations is to use a wet cure. *Wet cures* are brines in which the meat is submerged, which can be seasoned with herbs, spices, and other flavorings. At Larder we use this approach not just for cured meats that will be hung to dry but also for ones that will be cooked afterward, such as pastrami. This process adds large amounts of umami flavor that develops over the course of the cure. It doesn't have the intensity of the koji-cultured meats, but it still is a stunner. To create a koji brine, follow these steps:

1. Place the meat into a container or vacuum bag large enough to hold it.

2. Add enough amazake to fully submerge the meat. If you are using vacuum bags, you will need far less amazake than if you use a bucket.
3. Weigh the meat and amazake. Calculate 3 percent of its weight (minus the container weight) and add that much salt to the container or bag.
4. Allow the meat to fully cure based on its size and shape.
5. If the meat is to be cooked, such as pastrami or certain styles of ham, then do so. There is no need to rinse the amazake brine from the meat before cooking. In fact, we advise against it because it helps create so many delicious flavors as browning takes place.

The other way to cure meat with koji is to incorporate ground fresh or dried koji into a dry cure. As you have noticed throughout the book, we have been highlighting the work of folks who have been using koji in inspiring ways. In support of getting feedback from chefs known for being critical, we want to introduce you to our two friends who are always focused on flavor and quality, especially when it relates to cured meats. The first is David Uyger of Lucia in Dallas, Texas, whose knowledge of Italian cuisine, especially charcuterie, is unparalleled. The other is Nicco Muratore of Commonwealth in Cambridge, Massachusetts, who has culinary knowledge beyond his years and has been collaborating with Rich since he started on the koji journey. See their sidebars for their respective thoughts on and approaches to curing with koji (pages 198 and 200). Both of these chefs are highly regarded among their peers and receive acclaim for the foods that they create.

## Culturing Koji on Meat

Now that we've established how and why meats are aged, let's delve into how koji can age meat, or at least create meats with flavors similar to traditionally aged meats. We'll first walk you through the process and then discuss how and why this works. It is a straightforward culturing process and can be done in a wide variety of setups. The following tips will help ensure safe success:

## HONEST OPINIONS ON DRY CURING WITH KOJI

**David Uyger**

Admittedly most of my successful experiments involving koji have ended up being long-cured whole-muscle cuts. I've varied the cuts, but I primarily used pork, and found that the quality of the pork made a huge difference in the end result. For example: I cured some smaller Berkshire coppa that I got from a medium-sized pork producer. I'm not sure whether it was the high water content or the smallish size of the animal that produced a less-than-ideal result. They ended up being mushy and prone to exterior mold growth. That being said, coppa that I butchered from a large Red Wattle hog from a local farmer produced great results.

I suppose I'm getting ahead of myself and should really discuss the methods that I have used, or at least what worked and what didn't. I always used inoculated rice in the cure at standard refrigerator temperatures. I varied cuts, and I was generally able to figure out something that would work. I varied the amount of koji rice that I added to the cure. I varied the amount of water that was available during the curing phase (less is best here).

My method is pretty simple: I put appropriately cut and shaped whole muscle into an equilibrium cure for two to six weeks, rotating every few days. When it feels like the protein has denatured significantly and become firmer, I finish tying/shaping/stuffing into casing and hang until my desired doneness is reached (firmness and water loss being the main determining factors). The best results that I've had involved whole koji-inoculated rice. It would mostly break down during the curing phase, and I would scrape off the rest before drying. I think that this was mostly a textural preference.

As for the results, I enjoy occasionally using koji rice with *salumi*. The lean meat remains quite soft even after it has lost the appropriate amount of water. When it is done curing, the fat has more of a fudgy texture than standard salt curing without koji rice. The curing phase always seems faster with koji rice, however. As for flavor, any seasoning must be very strong to stand up to a cure containing rice. If I am curing something with the koji rice, I often double some seasonings and reapply if I want that flavor to stand out with the end product. This includes some pretty strong flavors like rosemary and juniper. Standard amounts of black pepper, rosemary, and bay are overwhelmed by the koji rice. That being said, the end result is really tasty! I would say that it tastes nutty, aged, and sweet. It makes the pork taste porkier in a very good way. I have found koji rice to be an interesting addition to the ways of flavoring salumi. It doesn't replace standard methods, but it is a nice variation.

- Stick with whole-muscle cuts of meats for this process. Don't use this technique on ground meats unless they are safely cured and the end goal is to create charcuterie. A large number of pathogenic microbes are surface dwelling and can be gotten rid of by cooking the cultured meats to the USDA-recommended safe internal temperatures, which is why you can eat a rare steak.
- Humidity should be maintained at 90 percent RH (relative humidity).
- Temperatures of 80 to 90°F (26–32°C) are ideal and should be maintained throughout the culturing process.
- Over-incubating can lead to spoilage. Most meats should need no more than thirty-six hours.
- Placing the meat on non-reactive resting racks (stainless steel or plastic are ideal), put inside a sheet tray, with minimal contact with the muscle, will ensure that you get a full culture.
- Adding a bit of water to the sheet tray ensures proper humidity.
- Work clean and don't cross-contaminate your work area.
- Various starches can be used to culture the meat. We like cornstarch best, but rice flour and tapioca starch work great, too.
- Cultured charcuterie can be hung immediately after culturing.
- Koji-aged meats should be chilled immediately after a full culture is established, and should be used or frozen within three days to prevent undesirable changes in the meat's texture due to moisture loss.
- Finally, the cultured meats can be cooked a variety of ways, from pan-fried to roasted to grilled.

## PRACTICAL CHARCUTERIE

**Nicco Muratore**

At Commonwealth we have the pleasure of working with beautiful heritage pigs. Working with the whole animal allows us to utilize cuts that are not always available to folks purchasing primal, subprimal, or market cuts. Before I started working with koji, we already had a small curing program in place. We would take the cheeks and cure them for guanciale, the back fat for lardo, the neck muscle for coppa, and so on.

One day Rich and I were making some guanciale, messing around with different cures for each one. He decided to add some koji to one of the cures to see what would happen. We already knew the enzymatic power of protease enzymes on protein substrates, so why not attempt to add the koji flavor bomb to our charcuterie? The result was an incredibly flavorful, unctuous, nutty, melt-in-your-mouth piece of cured meat. It also happened to slightly accelerate the drying process and yield a properly dried piece of cured meat (30 to 40 percent weight loss) in two-thirds of the time.

Traditional cured meats are already packed with umami. The protein is broken down by enzymes as the meat dries out, releasing the amino acids that we taste as umami. Using koji has allowed us to introduce these enzymes from the very beginning of the curing process, and elevate the flavor of the cured meat in a different way.

## NICCO'S KOJI COPPA

I love coppa, a cut of pork shoulder that is rich in fat. It is easy to butcher yourself and considered by many to be a great cut of meat to start with when making dry-cured meats. Measurements for salt and koji are by percentage weight of the whole coppa.

**1 pork coppa, trimmed and weighed (+/− 1 kg / 2.2 pounds)**
**3 percent kosher salt**
**0.25 percent pink salt #1 (Instacure #1, DQ curing salt #1)**
**5 percent koji, ground in a food processor or mortar and pestle**
**Spice mix (14 g [0.5 ounce] black pepper; 2 large sprigs fresh thyme or savory; 6 g [0.2 ounce] fennel seed; 1 bay leaf; 7 g [0.3 ounce] aleppo chili flakes; 3 cloves garlic, smashed)**

In a large bowl, measure out and mix all the ingredients, except for the coppa, into what constitutes the cure. Rub the cure onto all surfaces of the meat, massaging it lightly, using the salt as an abrasive to help get the koji equally distributed on the meat. Tightly wrap or vacuum-seal the meat. Let it cure in the fridge for 5 to 6 days, depending on size. Rinse the cure off under cold water and dry the meat. Pack into beef bung casings or wrap in cheesecloth, and tie the meat tightly. If you're using bung casings, poke holes with a small tack to allow moisture to leave the meat. Hang in either a curing chamber, a cool room, or the fridge until the meat has lost 30 to 40 percent of its initial weight.

The koji-cured coppa is a delicious cured meat, with elevated umami and a deepened flavor with trace notes of koji and slight sweetness. The fat melts in your mouth more so than a traditional cure. The texture tends to be slightly softer, but you can adjust that by drying out the meat closer to 40 percent moisture loss if you prefer a firmer end product.

# Koji-Cultured Steak

This recipe is an example of a koji meat-culturing setup that we use at Larder. It requires minimal investment, and all the components can be easily cleaned and sterilized. We outlined this method using New York strip steaks because they are widely accessible and delicious, and lead to what we feel is a shining example of what accelerated aging can accomplish. We've tested and applied this process to a wide variety of meats and seafood. In the end, our opinion is that the darker a meat, the more delicious. On that note we want to reiterate that poultry and seafood are very difficult to work with; we advise against using this technique with these unless you are comfortable with not just the culturing process but also identifying if something has gone wrong. There hasn't been any terrestrial meat that we've tested that didn't properly culture. That isn't to say that all meats cultured turned out to be delicious, however; they simply cultured well.

The first step of this process is to season the meat with salt and sugar to taste. (You would skip this step for charcuterie because the meat is already cured.) We do this for several reasons: for taste and flavor; to help ward off pathogens; and to bring moisture to the surface of the meat so that the cornstarch and koji spores can adhere. The sugar acts as instant food source for the fungus, which allows it to nourish itself as it grows faster than if it were to produce its enzymes then feed.

**2 New York strip steaks 1.2 to 2.5 cm thick (0.5–1 inch)**
**1.75 percent salt**
**1.75 percent sugar**
**3 g (0.1 ounce) *Aspergillus oryzae* (koji) spores, dispersed in rice flour**
**125 g (4.4 ounces) cornstarch**

Weigh the steaks. Calculate 1.75 percent of their weight, and in a large mixing bowl season the steaks with the calculated weight of salt and sugar. Let them sit for 20 minutes or until they start to weep their purge. Mix the spores and cornstarch together and then fully and evenly coat the steaks. Use more cornstarch as needed to ensure the meat is completely covered. Put the meat into the incubation setup. We often insert wire racks into a deep-sided tub and wrap the tub in plastic wrap. Culture the meat at 80 to 90°F (26–32°C) and 90 percent RH for 36 hours. Once the 36 hours has passed, the meat should be fully cultured. Reserve it in the refrigerator until you are ready to use.

## HOW DOES IT WORK?

As we discussed earlier, koji produces numerous enzymes in a process known as extracellular digestion. Specialized cells in the mold's hyphae and mycelium produce these enzymes to break down the fungi's food into easily absorbed nutrients. These nutrients are then directly absorbed in the mycelium and act to sustain the fungus. Koji's enzymes are especially powerful and fast acting and allow for the production of various base nutrients. Enzyme production and action is at its peak while the mold is growing. When koji grows on meat, its enzymes work not only on the starch coating but also on the proteins and fats in the meat. The meat is in such close proximity to the starch coating that these enzymes end up making direct contact with the surface. Sugars such as glucose and maltose, along with amino acids (namely glutamic), are produced during this breakdown. Essentially koji does the same work that the enzymes naturally found in meat do, but instead of taking a month or more, koji does this in less than forty-eight hours.

During our experimentation with and development of this technique, a question emerged as to whether this is the most efficient means to impart and optimize koji's enzymes on a piece of meat. The answer is no. Growing koji on a starch-dredged piece of meat isn't the most effective way to impart the enzymes and take advantage of their activity. The most effective way would be to use a syringe to *inject*, via shio koji, amazake, or purified enzymes, evenly into the meat and then optimize the temperature for them to act as catalysts. That being said, culturing the meat has turned out to be the most delicious method to use with meat. We have carried out numerous side-by-side taste tests of cultured meat next to not only dry- and wet-aged meats but also meats treated with a wide range of koji makes from amino pastes to shio koji.

# Beyond Whole-Muscle Applications

Dry-cured whole muscle isn't they only type of charcuterie that you can make using koji; in fact, many ground preparations can be made using koji. These can be cultured, but we prefer to use amazake in the grind. Pâté is a prime example of this; by using amazake in your pâté, you will get a better

## THINGS TO REMEMBER WHEN COOKING KOJI-AGED MEAT

There are several things to remember when you're cooking a koji-aged piece of meat:

- Leave the koji crust on. It creates the most amazing browning that you'll ever experience. You can remove it if you prefer, but we love how it eats.
- Speaking of browning, a koji-aged meat's will be greatly accelerated when cooking. Cook over low to medium-low heat or else you'll burn the meat on the outside.
- Be mindful that due to changes in the myofilaments of the meat, the scale of doneness has slid down. This means that the feel of a medium cook is actually going to now be closer to well done. It may take you a few tries to get comfortable with this.
- Wet-cooking methods such as braising can cause the koji crust to gel and develop an off-putting texture. Stick to dry methods like pan-frying, pan-searing, grilling, and roasting. If you're going to braise—and we often do—be sure to sear the crust first.
- Cooking the meat to a USDA-recommended safe temperature ensures that any potential surface-dwelling pathogens and anything that may be lurking inside are killed.

bind (the way the meat holds together) due to the residual starches from the rice and all the added flavor benefits that koji brings to the table.

*Pâté* translates from the French as "paste" but has come to reference what is essentially a refined version of meat loaf intended to be served cold. Making pâté is simple and can even be done with pre-ground meat from your butcher if you don't have a meat grinder. Traditional pâtés are often blends of pork, veal, and liver, but virtually any meat or meats can be used. The gefilte fish made from carp that Jeremy serves at Larder is a pâté. Pâté is considered an emulsified forcemeat and is a blend of muscle, fat, flour, eggs, spices and seasonings, and cream. A mixture of flour, eggs, and cream called a *panade* is mixed into the ground meat to ensure that it is bound to a fairly smooth texture in the finished cooked product.

## KOJI COMPARED WITH TRADITIONALLY AGED MEATS

**Diana Clark**

*How do we know that koji-aged can compete with traditional dry-aged meats? We tested it with Diana Clark, a scientist and bovine anatomist, who for many years has worked for Certified Angus Beef. While she finds differences between traditional dry-aged meats and koji-aged ones, we are unanimous in our love for the koji-aged, especially when the economics of aging are factored in. (This isn't to say we don't also love traditionally aged meats, because we do!)*

Have you ever made a wine reduction sauce? Actually, it is quite simple. Heat some butter and flour over medium heat, stir in wine, vinegar, and herbs, then simmer and reduce the volume of the sauce. Gradually, you are concentrating the flavors of the wine as water evaporates, allowing a small teaspoon to carry robust fruity flavors. The low and slow simmer of heat intensifies the flavors by creating new volatiles that change and enhance the flavor of the sauce.

Aging beef is a similar culinary work of art, in which beef is unconfined and exposed to the open air in a humidity- and temperature-controlled environment. A similar process, known as wet aging, involves holding the product in a sealed, vacuum package (air-removed bag) for an extended period of time. Both aging processes allow the natural enzymes within the meat to weaken and degrade proteins and, consequently, develop a more tender, flavorful product.

While subtle, the differences between wet and dry aging are critically important in that dry aging allows the product to be exposed to the environment, while wet aging contains the product within a sealed package. Because wet aging improves tenderness without dramatically reducing moisture levels and yield, it is common practice in the beef industry, such that the majority of beef consumed today is wet aged. While dry aging is less efficient, the benefits of this process center on flavor quality. Dry aging produces a robust, delicate, beefy flavor, which is sought out by many consumers who are willing to pay a little extra for an exceptional experience.

Standard methodology of dry aging is to keep beef at normal refrigeration temperature: between 29 and 39°F

(−2 to 4°C). Meat freezes at 28°F (−2°C), so temperatures below this would reduce or eliminate enzymatic activity and in essence inhibit the aging process. On the other hand, temperatures above 40°F (4°C) are a potential food safety risk. Relative humidity is commonly kept between 75 and 80 percent (+/− 5 percent). If humidity levels are too low, the meat can dry out before the unique flavors can develop, but if the humidity is too high the risk of rancidity increases. Last, but certainly not least, is airflow. The product must have consistent airflow to make the aging process uniform and balanced.

Interestingly, while the dry-aging flavor is often strong and desirable, we still do not understand how this flavor develops in beef. Some speculate that the dry-aging process results in the evaporation of water, which makes up approximately 70 to 75 percent of fresh beef. This reduction in water condenses and concentrates the "beefy" flavors much like the process of making a wine reduction sauce. Others attribute the unique flavor profile of dry-aged beef to molds developing on the surface of the meat.

As discussed previously in this book, mold can play an important part in food production, especially in cheese, soy, and some fermented products. In fact, specific strains of molds such as *Penicillium camemberti* are commonly used to ripen cheese by promoting the catabolism of lactose, lipid, and protein, thereby enhancing the flavor profile. These molds will break down components (lactose, lipids, or proteins) of cheese to form simpler molecules. Simplifying these molecules allows new aromas to develop. Have you ever tasted something when you have a cold? Most of the flavor is gone because your nose is stuffed up. Our olfactory glands have the ability to detect aromas and send them to our brain to "decode" the smells. Without our sense of smell, most foods taste the same. This is why subtle differences in molecules can make major impacts on food aromas and taste.

As koji catabolizes different starches and proteins, it creates smaller sugars and amino acids that positively influence the desired flavor profile of savory foods. This is the ultimate goal of dry aging beef: to condense the natural flavors of beef and intensify the nutty, rich, robust, beefy taste. When proper procedures are followed, koji should create the dry-aged flavor of forty-five-day-aged beefsteak in just a few days. The mold is thought to produce the desired simplified molecules that alter the overall beef aroma and taste. Koji has the potential to simplify the duration of the dry-aging process and deliver a more consistent product overall.

A basic ratio for a panade per pound (500 g) of ground meat is: 1 egg, 0.3 ounce (9 g) flour, and 0.5 ounce (14 g) heavy cream. If you're using amazake, simply replace the heavy cream with amazake. This would also apply to a mousseline or any other forcemeat that has a hydration. You can fully replace the hydration in any ground charcuterie preparation with amazake. You can also use other liquid koji-derived makes such as shio koji, amino sauces, and alcohol, but keep in mind the difference in salt, acid, and flavor that these will impart. The possibilities for working koji into your ground charcuterie are so vast that we consider them to be endless.

When you're seasoning a pâté, the salt should be kept to around 1.75 percent of the weight of all the other ingredients. We also recommend adding 0.25 percent of Cure #1 to keep the pâté from turning a disgusting gray color. (Substitute celery powder or juice should you prefer not to use curing salts.) You can garnish the pâté with a wide variety of cooked meats or fats, cooked vegetables or mushrooms, nuts and dried fruits, or a recent favorite of ours: douchi. This garnish is folded into the meat after the meat is ground. Cooking is key here. The size, density, and shape of your garnish is vastly different from the rest of the grind. To avoid over- or undercooking your garnish, always precook and then chill it before adding it to your raw grind. You can use grinds of different coarseness to give your pâté more rustic textures and a spectacular visual appearance. As with the cured meats we discussed above, spices and seasoning should be added to taste and personal preference. This includes seasonings such as brandy.

Pâté is typically cooked slowly at low temperatures. The mixture is placed into a loaf pan and cooked either sous-vide or in a low oven, 250°F (120°C), in a pan of water until the internal temperature is 165°F (73°C). The pan of water ensures that the pâté never reaches a temperature above the boiling point of water. Doing so will overcook your pâté, leaving it with a grainy texture. Use a thermometer, inserted into the center of the pâté, to determine when it is done cooking. Allow your pâté to rest at least overnight in the refrigerator. Most will hold for up to two weeks under refrigeration.

In addition, dry-cured ground styles of charcuterie such as *landjaeger* and *sopressata* can also be made with koji. There are different routes you can take, such as culturing the charcuterie or incorporating koji into the grind as the hydration. As we stated previously you can also simply take your favorite recipe and substitute amazake for the water called for in the salami. Keep in mind that you will have to experiment and examine how your new recipe acts to ensure that it is not only delicious but also safe.

For these styles we recommend using an amazake that has been made from protease-heavy koji. This keeps the levels of sugars that you are adding to the sausage to a minimum and should not create any food safety issues. The addition of sugars from an amazake can cause differences in the fermentation of some of these styles of meat. Always test the pH and $a_w$ of these meats to ensure they are safe before consuming.

Generally speaking, when making charcuterie, such as salami, from ground meat, we prefer adding amazake to a grind instead of culturing the charcuterie. The main reason for this is that koji's growth is so vigorous and active, it can draw too much water too quickly out of the grind if it has been cultured on the outside. This can lead to grain separation and gapping inside the cased charcuterie and can increase the risk of pathogenic contamination, including botulinum contamination and poisoning.

At Larder we make a mettwurst sausage that has been a mainstay of our charcuterie board almost since we opened. *Mettwurst* is a large family of fermented, spreadable sausages that are made throughout Germany. It is in the same family of sausages as the Italian *n'duja* and the Spanish *sobrasada*. Instead of using water in this sausage, we add amazake. We use 2 cups (473 ml) of amazake per 10 pounds (4.5 kg) of meat. This ratio works great and is somewhat standard in many dry-cured ground charcuterie recipes.

## Vegetables That Eat Like Cured Meat

While the majority of this chapter has been devoted to age-accelerating meat and the making of charcuterie using koji, we would be remiss if we didn't talk about how all these techniques can be applied to vegetables and even fruits. While they don't contain large amounts, vegetables and fruits do contain protein, along with fats and the obvious carbohydrates, and can act as the perfect medium for koji to work its magic on. From optimized marination to culturing, you can transform your vegetables into something captivating and spellbinding for your diners. You can even create vegetable-based charcuterie using these methods.

For many years people adhering to vegetarian and vegan diets have sought foods that replace the nutritive value of animal proteins as well as satiating ingrained primal urges to eat meat. This is evident by the sheer number of meat alternatives that are available to us, from traditional Asian foods like seitan and tofu to modern Western ones like fake chicken nuggets and the incredibly fascinating Impossible Burger. In our quest to create vegetable charcuterie, we wanted to mimic the flavor, texture, and visual appeal of meat-based charcuterie. Over the past couple of years, we have developed techniques to make this vegetable-based charcuterie. At their root the techniques aren't too different from the above-outlined ones for whole-muscle charcuterie. Before we dive into how to create these foods, however, we want to discuss why we refer to these as charcuterie.

Purists will insist that charcuterie can only be made from animal tissue. For most the term can include oceanic proteins such as lox or a bresaola made from tuna or swordfish, but fervent purists consider even these blasphemous, let alone charcuterie made from a vegetable. But we define charcuterie by the technique and methodology behind its production rather than the ingredient it's made from. These techniques primarily stick to salting, or curing, and drying but can also include various cooking methods such as poaching and hot smoking. We feel that under our definition any ingredient can be used to make charcuterie and have chosen to label these vegetable-based ones as such. Many individuals who have argued this with us haven't provided us with alternative names by which to call these foods, so *vegetable charcuterie* is what we've stuck with. These are vegetables that have been cured, inoculated with mold, and then hung to dry. If a piece of meat went through these steps we would call it charcuterie, so why wouldn't that extend to a vegetable?

When making charcuterie from a vegetable, it is firstly important to realize how it is different from animal tissue. While it is still made of the same few core building blocks that make up all life on Earth, they are found in not only different combinations but different ratios as well. This being the case one's thought process has to wander into the realm of textural transformation and how that can be achieved. The transformation of taste is easy enough to imagine; with enough salt, smoke, spices, and our friend the Maillard reaction, most fruits and vegetables can be made to taste like a piece of meat. What really needs to happen is a textural transformation so that the vegetable charcuterie not only looks like meat but also feels like it on our tongues. This all starts with treating a vegetable as if it were meat.

Choosing the proper vegetable or fruit is paramount. Soft fruits or vegetables such as a tomato or a peach are too delicate to transform into a piece of vegetable charcuterie with a meaty texture. Choose something firmer such as a beet. Stay away from foods with potentially harsh pungency such as turnips, or overtly sweet ones such as plums. Also stay away from vegetables that don't have any bulk or body to them, such as kale and other leafy greens. Here are some vegetables and fruits that can work well with our techniques:

- Beets
- Carrots
- Broccoli, especially the peeled stems
- Various winter squash, such as spaghetti and Hubbard
- Summer squash such as zucchini and crookneck
- Daikon and other mild radishes
- Firm apples and pears
- Burdock root
- Chinese long beans
- Firm mushrooms such as maitake (*Grifola frondosa*) or chicken of the woods (*Laetiporus sulphureus*)

These suggestions are in no particular order, and all work well. We do, however, enjoy beets, carrots, Chinese long beans, and maitake the best.

The first thing that needs to be achieved is to soften the vegetable's texture. This is best done by either hot smoking or boiling. If the texture of the vegetable isn't changed before curing and inoculation, it runs the risk of staying crisp and crunchy, which isn't what we're going for here. Multiple cooking techniques can be used should you choose. The beet charcuterie that we produce is first boiled and then smoked.

It should be noted that vegetables cured to the 3 percent used in the EQ method end up being way too salty by the time they are dried. This being so, we only cure them to 1.75 percent. This just so happens to be the ratio of salt that we use for charcuterie such as pâté and fresh sausages. While virtually any species or variant of *Aspergillus* will work, we like to use BF1 or BF2 from Higuchi Matsunosuke Shoten Co. Ltd. The Light Rice Miso spores from GEM Cultures also work well for this technique.

Here are the steps that we use to produce our beet vegetable charcuterie:

1. Boil the beets until they are fully cooked. Peel the beets after boiling.
2. Smoke the beets in hot smoke for 30 to 60 minutes.
3. Let the beets cool to room temperature and then rub them with salt. They can be sealed in a vacuum bag or wrapped tightly in plastic. Spices and other seasonings can be added at this time if you desire.
4. Allow the beets to cure under refrigeration for at least 2 days for baseball-sized ones and a few days longer for exceedingly large ones. We've never had beets overcure, so leaving them for up to 10 days is fine.
5. Remove the beets from the cure and gently pat them dry.
6. Use a very fine-mesh strainer to evenly sprinkle dispersed spores onto the surface of the beets.
7. There is no need to add excess starch here as we do for meat, because the beet has plenty of starch for the koji to access.
8. Place the beets in your setup for meat and culture them the same way.
9. Once the koji has fully covered the beet and bloomed, it is ready to dry.
10. Weigh the cultured beets and log their weight.
11. You can either tie the beets up and hang them in your charcuterie chamber or dry them in a dehydrator at 90°F (32°C). They are done when they lose roughly 50 to 60 percent of their starting weight.

This process works broadly with any of the fruits and vegetables that we mentioned and allows for differences in flavor to be achieved depending on your initial cooking method, the spices and seasonings you use, and whether or not you choose to smoke them. You can also ferment these foods before culturing to give them flavor profiles often found in styles of ground charcuterie. When finished you will end up with a salty, umami-laden product that is remarkably reminiscent of styles of whole-muscle charcuterie. Simply slice it thin and enjoy it as part of a charcuterie board.

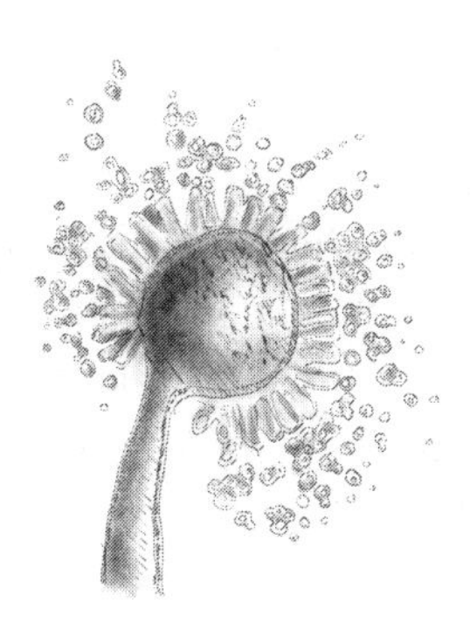

## CHAPTER 11

# Dairy and Eggs

Dairy products and eggs are packed with protein and have serious umami potential. They're nutritionally dense, so it makes sense that they're key components of cuisines all over the world. It doesn't hurt that they're also delicious. Most traditional preparations involve preservation of some form as they do not last long sitting out on their own. Baking is a short-term obvious choice. However, salting and aging long ago arose as the best option. Think cheese.

One of the best ways to preserve anything is to concentrate and salt it. However, making cheese isn't as simple as simmering milk on a stove and adding salt. There are important pieces to the process that make it work, and the end result ultimately amazing. One of them is *rennet*, enzymes from a ruminant's stomach used to coagulate milk so it can be formed into a solid. The other is the introduction of microbes at different stages. Originally these microbes came from the environment; they're now specifically selected for the numerous styles as well as product consistency. Most important, microbes create conditions to make the food safe and delicious. As part of the process, they also bring along all the wonderful enzymes we've been talking about. Cheese is not much different than miso in that both are amazing foods drived from a nutritionally rich, humble product. The only hitch is that the microbes on cheese aren't nearly as efficient as koji is at producing enzymes.

By applying koji to dairy and eggs, you can realize a complexity of flavor in a matter of months. This is especially amazing when it comes to dairy proteins. With koji, aged cheese flavors develop on the order of one-quarter the time you'd normally expect. Techniques to make these delicious foods are as straightforward as their traditional equivalents supplemented with a percentage of koji depending upon the application. We have selected methods and recipes to yield a wide range of delicious products to add to your pantry. As you will see, these makes are open to milk and eggs from pretty much any animal in the world that is consumed for nutrition.

## A NEAT WAY TO ADDRESS INADVERTENT ALCOHOL BUILDUP

When fermenting amino pastes and amino sauces, don't air lock or seal them—doing so has the tendency to retain alcohol that's produced. Instead you want to let this evaporate, as it isn't a favorable addition to the flavor. However, sometimes you end up not having time to maintain the product appropriately and it results in that undesirable flavor. We've found that even though it isn't ideal, the product you have is still pretty good. All you have to do is find a way to eliminate the alcohol. Simply heat it lightly to evaporate the alcohol.

# Amazake-Cultured Cream

One of the first dairy ferments discovered was most certainly cultured cream. Imagine a day thousands of years ago when a fresh bucket of milk was left out too long, providing a perfect environment for *Lactobacillus* and other nutrition-seeking microbes, which in turn soured the cream such that it could be held for a little longer and made it desirable in a different way. Then the cream top could be taken and beaten into butter and salted so it could be preserved.

When it comes to making thick and luxurious cultured cream, choose whatever minimally processed cream you feel comfortable consuming. Raw milk cultured cream is quite the treat if you're open to that. Needless to say, standard high-volume-produced cartons work well, but the cream doesn't come out nearly as thick or taste as complex.

Amazake-cultured cream is essentially a koji-based sour cream. Once the cream is made, it can be churned into a butter and buttermilk that are transcendent in their flavor and use. It has an amazing ability to resist curdling and separating when heated. Jeremy's co-owner at Larder, the venerable Kenny Scott, says he can literally boil this cultured cream and it stays just as thick as if it were fresh. He says it is the perfect dairy culture with which to mount pan sauces and thicken creamy soups.

**2 cups (500 g) heavy cream**
**1 ounce (30 g) amazake**

Mix the heavy cream and amazake together in a non-reactive bowl or vessel. Cover with a kitchen towel or cheesecloth. Allow it to sit at ambient temperature until it significantly thickens and has a sour note. This usually takes about 24 hours.

# Miso Cheese

One of the first things we used as a guide for making off-the-beaten-path amino pastes and sauces, and still turn to today, was the product nutrition label that can be found on nearly any packaged food. This made a quick reference for understanding exact protein, carbohydrate, and fat content in most foods. Armed with this understanding, we could choose alternative products that made logical sense. How close was what we were making to what's used in a commercial amino paste? Of course, we wanted to try something completely new that didn't fall into the legume, nut, or seed category. We eventually decided upon ricotta cheese. It has a fair percentage of protein, slightly above typical beans, along with much lower carbohydrates and a fair bit more fat. For sure, it was close enough to give it a go.

Making an amino paste with fresh cheese was one of our first breakthroughs at the beginning of our investigations. Applying koji enzymes to dairy proteins yields aged cheese flavors as soon as two months, which is five times faster than Parmesan. Unparalleled potential!

**250 g koji (fresh preferred)**
**25 g kosher salt**
**250 g ricotta or other fresh cheese**

Add the koji and salt to a medium mixing bowl. Note: If you're using dry koji, in a small bowl mix 25 grams of lukewarm water with 225 grams of koji and allow it to hydrate for a couple of hours at room temperature. If you don't want to wait, process into a rough paste.

With clean hands, mix the koji and salt together so they're evenly distributed. Now combine mixing and squeezing the koji and salt together to break down the koji into a paste as much as possible. Don't worry too much about making it super fine or if you miss some grains. The pieces will have the opportunity to break down during the fermentation process. It also can easily be blended afterward to make it smooth.

Add the ricotta and mix thoroughly. Pour the contents into any non-reactive pint container. A pint mason jar is preferred, but a deli container will work just as well. Cover the mixture's surface with plastic wrap and cover with a lid. If you're using a mason jar, do not seal. Place the jar in your refrigerator for at least 2 months. After the time is up, taste the Miso Cheese. If it has the flavor of an aged cheese similar to Parmesan or Romano, it's done. If not, put it back in the refrigerator for another month until the flavor comes through.

At this moment, you might be asking why are you fermenting in the refrigerator. Ricotta inherently has a high fat content, and the warmer temperatures that we recommended earlier may cause rancidity. We've found that the controlled refrigerator temperature still allows enough enzyme activity to break proteins down into tasty amino acids. This is especially helpful in commercial kitchens, where a controlled environment is hard to come by.

Once it's done, use this preparation as you would miso or an aged cheese you'd add to a recipe. One simple and delicious application is to make a compound butter by mixing 1 tablespoon of Miso Cheese into a stick of butter. There's also the option of drying it so it's more like a hard cheese. All you have to do is put the miso paste in the middle of four layers of cheesecloth, squeeze the liquid tamari out into a bowl (save it to use in place of soy sauce), flatten the cheese into a 0.5-inch-thick (1.2 cm) disk, then hang it to dry, preferably between 40 and 50°F (4–10°C) in a low-humidity environment.

# James Wayman's Caramelized Miso Cheese

Our good friend James Wayman, chef of the Oyster Club in Mystic, Connecticut, is making a huge impact on the southern New England food scene. He was inspired by our Miso Cheese and decided to make his own. As a result, he came up with a variation that we had to share.

**453 g goat cheese**
**226 g jasmine rice koji**
**33 g sea salt**

As you may have noticed, this recipe calls for a 2:1 ratio of protein base to koji. This ultimately increases the protein-to-starch ratio, which leans toward more umami (with less sugar for sweetness) as well as ultimate acidity in the fermentation, and follows along the lines of a traditional long-term miso that's focused on protein conversion for umami over feeding microbes.

Follow the mixing and containment instructions in the recipe for Miso Cheese. Let the cheese do its thing at ambient temperatures for a month. (You'll have to keep your eye out for the possibility of rancidity, of course. However, in all of our years of making Miso Cheese at ambient temperatures, we've never had an issue with it tasting bad.)

After the set duration, form a ball with the miso and hang it in cheesecloth for another week at ambient temperatures. Due to the ball shape, a touch of alcohol will build up inside. Break up the ball in a medium-sized bowl so the largest pieces are no bigger than 0.5 inch (1.2 cm). Spread it evenly on trays in a dehydrator set for 140°F (60°C) and run it for 8 hours. You'll end up with an amazing caramelized crumble that has crazy umami to top whatever you'd normally grate a sharp cheese or Parmesan over.

As an extension of this technique, you may be thinking that smoked cheese as a base for making an amino paste would be a good way to go. Well, we've found that there are times when it doesn't work out well. Smoking generates creosote, which is deposited on the food. It's acrid, bitter, and binds well to fats. We've found that when you use smoked cheese as a base for a miso, the undesirable flavor overpowers. This is especially true when you dehydrate it. So if you would like to make a smoked version, we recommend doing a cold smoke after all is said and done.

## Yogurt Miso Hot Sauce

We developed this hot sauce as an extension of the practice of adding a starter culture to a fermented hot sauce. One of the recommendations is to use yogurt to bring that specific *Lactobacillus* combination to the party. Based on our success with ricotta miso, we thought we'd get hit of cheesy umami if we added koji along with the yogurt in a hot sauce application. It worked very well. One of the best compliments that we got was that one of the hot sauces tasted like jalapeño poppers. To make a yogurt miso hot sauce, all you need to do is blend by weight 3 parts hot peppers, 1 part koji, and 1 part yogurt plus 5 percent salt against the total. Contain as you would for any lacto-ferment and let it sit at ambient temperatures for a week or so until it tastes like the best cheesy hot sauce you've ever had.

# Larder Black Cheese

This cheese is one that Jeremy makes at Larder using cream cheese and black koji rice. It is served as one of the cheese choices on the charcuterie board and is often used on roast beef sandwiches or paired with the whitefish or steelhead caviar that they make fresh. Black koji rice, cream cheese, and salt are mixed and allowed to ferment for several days. The result is a spreadable cheese that eats like a fine blue cheese. The amount of black koji rice called for in this recipe is 30 percent of the weight of the cream cheese; the salt is 3 percent of the weight of the cream cheese and the koji rice combined. Based on this ratio you can scale this up or down very easily. Should you not have *Aspergillus luchuensis* var. *awamori*, you can use brown koji, *A. luchuensis*. The citric acid produced by these strains is part of what makes the cheese so delicious.

**500 g cream cheese, softened**
**150 g black koji (*Aspergillu luchuensis* var. *awamori*) rice, fresh not dried**
**20 g salt**

Combine all the ingredients in a stand mixer and mix on medium speed for a few minutes or until it's uniform in color. Seal into a vacuum bag and let sit at ambient temperatures for 3 to 5 days. Remove the black cheese from the bag and store refrigerated until you need to use it. The cheese will have a shelf life of 4 to 6 weeks refrigerated but can be formed into bricks and dried to be used as a hard cheese, grated on anything from pasta to salads.

The two big themes of this book are interchangeability of base ingredients and no waste. What we said previously about Miso Cheese doesn't just apply to fresh cheese; it can be applied to finished cheese of any kind. With that in mind, there are those times when you have a large amount of cut cheese sitting around after an event, or odds and ends you've been saving for a cheese sauce. This can easily turn into an even more complex cheese that you've never tasted before. The purpose of an amino paste is to maximize umami with all of the koji enzymes available. Why not "waste not" in the most delicious way possible?

# Whey Liquid Amino

Sam Jett once again lends his expertise and tells us that whenever he's producing ricotta, there's plenty of whey to be had. Armed with koji, making an amino sauce with this liquid was inevitable. How does it taste? Just imagine the marriage of soy sauce and the cheese powder found in your favorite mac-and-cheese box.

The key to this specific make is accelerating the process by leveraging heat. Holding the ingredients at an elevated temperature maximizes protease enzyme activity to ultimately yield a flavorful product in a much shorter time than a traditional soy sauce requires.

**3,785 g buttermilk whey (any whey will do honestly)**
**379 g uncooked rice, charred very heavily over embers**
**833 g koji**
**400 g salt**

Mix all the ingredients together and place in a non-reactive vessel. Wrap with a heating blanket or otherwise hold at 140°F (60°C). Secure an air lock or wrap a lid tightly to avoid moisture loss while being held at an elevated temperature. Stir daily for the first 7 days to ensure even heating. Let it go until you enjoy the flavor. Typically, Sam pulls it at 30 days for the optimal cheesiness and at 60 for a deeper soy-sauce-like flavor. Strain and bottle.

## CULTURES AND ENZYMES DRIVE CHEESEMAKING

Cultures and enzymes give us the ability to preserve, ferment, and flavor cheeses, and specific microbes along with time and temperature lead to the wide variety of products being made today. An important thing to understand is that the cultures used in cheesemaking generate enzymes to create complex flavors, similarly to how koji works, but not nearly as concentrated. These days cheesemaking is so standardized that most microbes and enzymes are created in laboratories for predictable results. We must keep in mind and not overlook the fact that cheesemaking started long ago with the intention of making a nutritionally rich preserved product. At the time, folks only had basic tools, limited resources, and naturally occurring cultures. This required a level of flexibility and ultimately led to folks around the world independently figuring out the variety of processes we see today. If we were to extend that spirit to now with all that is available to us, shouldn't that lead to more delicious cheeses?

Koji cheese is an interesting adventure. If cheesemakers are already using enzymes and cultures to make cheese, how is that any different from using koji? Well, it's not. The only big difference is the enzyme loading, as we've already seen in the Miso Cheese and amino sauces. Applying the enzymes produced by *Aspergillus oryzae* to cheesemaking processes has unparalleled potential. Of course, this is not a new idea, as enzymes have been industrially produced for their functionality in food for a long time. The key to all of this is that koji making is an efficient producer of food enzymes that anyone can do very simply and inexpensively. Another avenue to pursue is that of a holistic approach. This is still in its infancy at the time of this writing but is starting to be worked on by Kristyn and Kevin Henslee. Throughout all our experimentation, we've always wanted to seriously investigate what's possible with cheesemaking. Our only limitation was the lack of experience. As resourceful as we are in finding the right people, we were fortunate enough to befriend the owners of a local dairy farm who were more than willing to support our curiosity. Kevin and Kristyn Henslee own and operate Yellow House Cheese in Seville, Ohio. It's a small family farm that supports a flock of dairy sheep and makes award-winning farmstead cheeses. Their mission is to produce high-quality, handmade, small-batch cheeses, focusing on high standards of animal care while upholding the tradition of the family farm in local agriculture. Eventually, they attended a koji class. As a science educator, Kevin especially enjoyed learning more about how *A. oryzae* can transform food. This led to everyone's mutual interest in seeing how this could be applied to cheese.

The approach we speak of is to not only use koji enzymes to create better cheeses but also use citric-acid-producing species of *Aspergillus* such as *A. luchuensis* and *A. luchuensis* var. *awamori* to actually replace the rennet used to curdle the milk. Acids can act to curdle milk, as is done in the production of mozzarella or ricotta, and when paired with the enzymes present in koji they could act harmoniously and holistically as a one-stop cheesemaking tool. Expound on that and use a species of *Aspergillus* to culture the cheese and grow either within it or on its surface and you truly have something revolutionary. Now, that's empowering.

So where do we start? Well, we asked the expert and Kristyn decided upon a washed-rind cheese as it would be the easiest to incorporate koji. A washed-rind cheese is typically dipped in a brine solution so the surface offers favorable conditions for cultures to take hold during the aging process. Therefore it only made sense to start with shio koji, conveniently a brine in itself. In the initial experimentation, Kristyn washed cheeses in the umami magic brine. She found that it produced the funky aroma you traditionally smell. Upon tasting, Kevin picked up flavors that he has never tasted before and found it to be delicious. It was difficult for him to describe, and he's sure that it was the koji imparting new flavors and flavor dynamics to the cheese. This success has led the couple to continue to work on realizing the potential of koji cheeses. Kristyn is most excited about the culture that will result as the cheese is aged. We can't wait to taste the results. More important, we hope this plants a seed of change in how cheesemakers see their processes and how they can be transformed by a single ingredient. (Coincidentally on the other side of the world, we recently learned of two cheesemakers in Japan who came to the same independent conclusion as we did. They are using various koji makes to wash the European-style cheeses they make and are have great success.)

After Kristyn's success with using shio koji, she started testing the washing process with amazake. This happened to yield an even more superior product! Kristyn theorizes that the salt in the shio was actually acting to inhibit some of the molds and bacteria that would normally culture a washed-rind cheese, while the sugars in the amazake worked to feed these cultures. This further invigorated Kristyn to explore more ways to use koji in cheesemaking. The most fascinating of these was to add koji to the milk as she cooks and curdles it. She puts fresh koji into a sachet and cooks it into the milk. After the curd has started to form, Kristyn squeezes the sachet to extract every last bit of the enzymes present. From there the curd is cut, salted, drained, and packed into molds. The cheese is then aged to allow it to dry and the curds to meld together. This drying process, known as aging, typically takes a few months for the style of cheeses Kristyn makes. A cheese that has had koji steeped into it is perfectly aged in three weeks. At the three-week mark this yet-to-be-named cheese eats as if it were a twelve-month-old cheddar.

This experiment has been so successful that Kristyn and Kevin have gone into full production of koji cheeses. They can now be found at Larder and at the farmers markets in the Cleveland area that they attend every week to sell their products. As far as we're concerned, this could be the biggest advancement in cheesemaking since the advent of purified rennet. The economic impact for the cheesemaker who can create a cheese in three weeks that tastes and eats just as good (and in our opinion better) than ones that are aged for months to years is substantial. Imagine being able to sell your product nearly as fast as you make it instead of waiting for what can seem like ages.

As time goes on we hope that the methods and techniques that Kristyn and Kevin are pioneering at Yellow House Cheese will be nothing short of revolutionary and eventually widely implemented. Koji and its enzymes have once again shown us that the possibilities to make our food taste better, as well as to make its production more economically feasible, are virtually endless. All it requires are a few willing individuals who want to experiment and then share their ideas.

# Koji-Cured Egg Yolk

Cured egg yolks are an easy and wonderful way to bring not only a touch of elegance but also a serious, rich, eggy umami flavor to a dish. The yolks can be grated on everything from mac-and-cheese to vanilla ice cream. The general idea of doing this was borrowed from the Sardinian tradition of making *bottarga*, the salted and dried egg sacs of either tuna or mullet. Bottarga is grated over everything from dressed greens to pasta to give the dishes an umami boost. We've seen poultry eggs preserved and used this way and have been using them this way ourselves for many years now. When we first started making these yolks, we solely used salt to preserve them, but as time went on and we became passionate about koji, it made perfect sense to use koji or one of its makes to make these yolks.

**12 egg yolks**
**Shio koji, enough to cover the yolks**

Submerge the egg yolks in the shio koji. Leave them in the shio for 2 hours. Then remove the yolks from the shio and allow them to slowly dry in a dehydrator at 80 to 90°F (26–32°C) for at least a few hours and up to 1 day.

If you want, you can form the egg yolks into a log by gently pressing them together about halfway through the drying time. These logs can be shaved instead of grated and you'll obtain sheets of cured yolk that look great garnishing any dish. A friend of ours, chef Matt Danko, formerly of Grace in Chicago, Illinois, would even take slices of black truffles and sandwich them in between each yolk before molding the log.

# Koji Bottarga

The bottarga that Jeremy makes at Larder is really no different from the poultry yolk version, except that it's made from fish roe sacs. At Larder they only source fish out of the Great Lakes and freshwater rivers surrounding Cleveland, and every spring when the walleye start to run, they become plentiful beyond imagination. During this time the fish are packed with sperm and eggs. Jeremy works really close with his local fishmonger, Tom McIntyre, to source as many of the egg sacs that he can, sustainably, in order to make bottarga. Two things to note when sourcing fish eggs: You need the whole sac to be mainly intact, and you need small eggs. Eggs that come from whitefish, trout, salmon, and sturgeon should be saved to make caviar, not bottarga. The eggs of walleye, various bass, perch, and smelt all work great for bottarga.

**12 fish egg sacs**
**Shio koji, enough to cover the egg sacs**

Submerge the egg sacs into the shio koji. Leave them in the shio for at least 2 hours and up to overnight if the sacs are large. Some egg sacs can be as long and as big around as your forearm and will need to spend a longer time in the shio koji. Remove the egg sacs from the shio and allow them to slowly dry in a dehydrator at 80 to 90°F (26–32°C) for at least a few hours and up to 1 day. The bottarga can dry out if not kept tightly wrapped after it is made. Store it in a vacuum-sealed bag—or better yet, dip it in melted beeswax and it'll always be perfectly fudgy.

# Fish Roe Amino Paste

The first fish roe we tried was sweetfish roe—what you typically experience as the pleasant tasty pops on sushi rolls. We were enamored with the idea of being able to have a shelf-stable fish sauce in the same form. We figured that if we were able to leverage the salt in the mix to draw out the roe liquid so it would get infused with koji then migrate back into the roe, that would be amazing. However, we weren't sure if the roe would end up getting consumed by enzymes and dissolve. Interestingly enough, it was successful, making roe that popped with a BBQ fish sauce flavor after a few months and added a touch of depth for up to a year, and hasn't changed much since then. Since then we have gone on to make this product with not just various fish roe but with the sperm sacs, also known as soft roe, as well.

This is the ultimate seafood condiment that we find ourselves reaching for whenever we see a recipe call for anchovies or other fishy umami boosters. A favorite way for us to use it is to schmear some on a piece of buttered toast with thin slices of pickled onion, slices of ripe summer tomatoes, and chopped herbs. To make this we use a ratio of 1:4 koji to eggs or sperm. We then heavily salt the koji and eggs/sperm to 10 percent of their combined weight and let this autolyse for about a month. Age it as you would any other amino paste, as outlined in chapter 7.

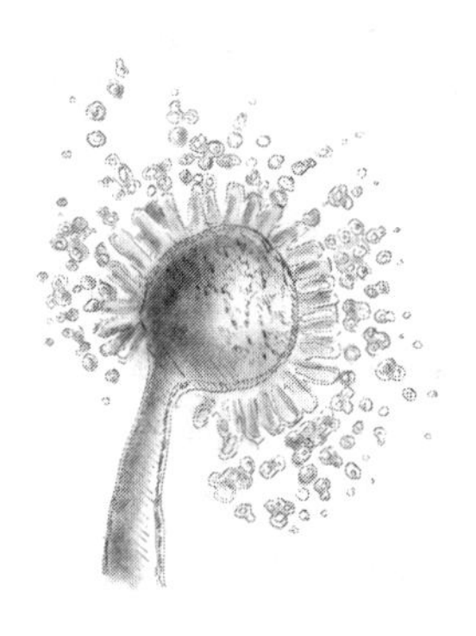

CHAPTER 12

# Vegetables

When it comes to vegetables, leveraging koji is a wonderful way to develop sugars to power different stages of pickles over time. Of course, koji also brings a touch of umami. Applications range from a short-term sweet-and-sour lacto-fermented pickle to a long-term aged *takuan*. When it comes down to getting started, just add koji to any pickling preparation. Is it really that simple? Well, it's important to have a fundamental understanding of the process and what makes sense to try. As we've previously discussed, koji is an active culture that you can use in place of any naturally started fermentation process. The same applies here. The important thing to realize is that if you load your ferment with koji, the enzymes will mess with the texture of most common pickling vegetables. We've found that the best ways to manage this are to add koji sparingly or to use the ferment as an accent to be served within a few hours. Of course, you can use the any of the delicious products of koji to pickle with, rice wine vinegar being an obvious choice.

Above and beyond that, there are also options for ingredients as a result of the "use every last bit" principle we have been discussing throughout this book. There is so much depth of flavor in what remains after you've harnessed the primary product. As an obvious extension of application, the pickling bases/brines can be used in place of the combination of citrus/acid and salt in cooking, as they're not far from a vinaigrette or marinade. Consider the by-product of sake fermentation commonly known as kasuzuke (sake lees pickles). Similarly, the solids from pressing out soy sauce are pretty much a diluted miso. Venturing out even further beyond using everything from a successful ferment, consider the failures as well. Especially with koji pickling, you will likely have applications that go too far: The texture isn't right, the vegetable has a strong alcohol flavor, you have tough scraps from root vegetables you pickled whole, or what have you. A lot of times the flavor can be concentrated and made quite good by simply dehydrating the solid matter then grinding it to be used as a seasoning. It becomes especially wonderful when you can use that lacto-salt to season the same vegetable.

Before Rich knew anything about koji, it was already a part of his childhood food experiences. One of his earliest memories was a refrigerator pickle his mom would make in the summer. It consisted of slices of cucumber thrown into a mix of water, soy sauce, rice wine vinegar, sambal (chili paste), and a touch of sugar. As always, the brine was made to taste. It was a quick and easy preparation that resulted in a refreshing delight that he'd reach for on a hot day. The beauty of the process was that there would always be different levels of crunch depending upon how long the cucumbers sat in the brine. To this day, he still

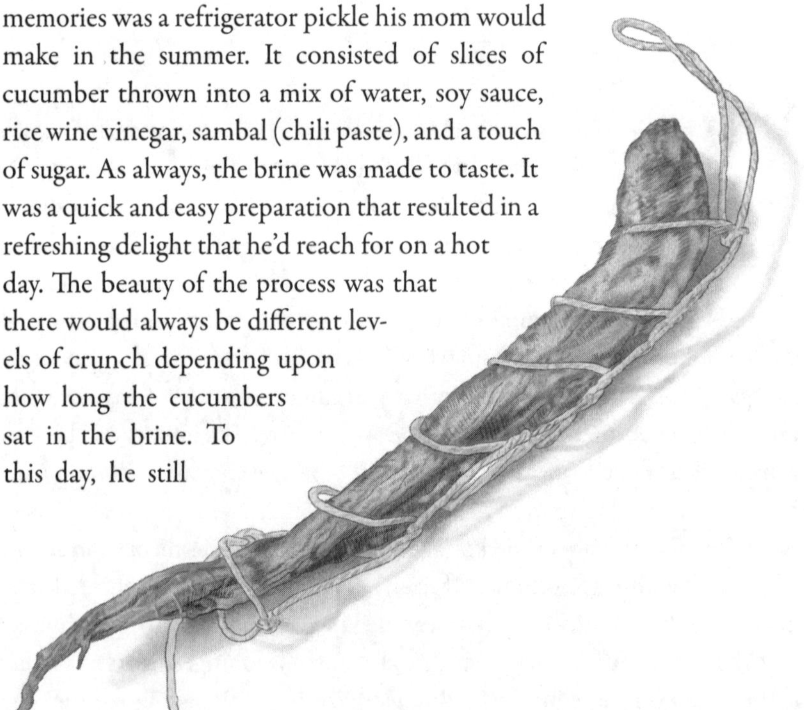

Air-dried daikon kojizuke. *Illustration by Max Hull.*

makes his own. However, he does take liberties by using a combination of the wide range of ferments he has on tap.

One of our favorite pickling applications to play around with is kojizuke. We discovered this technique in *The Book of Miso* specifically as bettarazuke, and the method we developed is inspired by it. Traditionally it's daikon pickled in a fermented slurry that consists of cooked rice, rice koji, sake, and salt. Prior to pickling, the daikon is air-dried and salt-pressed. This creates a very specific structure and gives the vegetable a unique crunch that can be sustained during the pickling process.

As we've discussed before with other processes, the flexibility of pickling is no different; each ingredient can be interchanged with whatever fits into the general category. Don't limit yourself to rice and sake. As much as the flavors of the other makes we've talked about have been driven by the base ingredients, it's especially important in this application as there is a shorter time to influence the final product. The pickling slurry is prepared by mixing 4 parts cooked grain, 4 parts koji, 1 part alcohol (with an ABV in the range of 5 to 15 percent). We've found that white wine, beer, and hard cider all work great. Add 5 percent salt against the total weight. Let the mix sit at ambient temperatures for a few days in a jar loosely covered with the lid, a kitchen towel, or cheesecloth rubber-banded over the top. This allows the amylase enzymes to further break down the starches into sugars in conjunction with the lacto-fermentation that adds a tart note.

Why use koji here when you could just use one of the umami-hopped-up amino sauces or pastes along with a custom vinegar to pickle with? That would be delicious, for sure! We can't argue that. However, this is an instance when koji elevates flavors ever so slightly so as not to overwhelm the base product. Leveraging the subtleties of umami and sugars developed from the vegetable will allow it to shine.

Once the pickling medium is ready, all you do is put vegetables in contact with it and enjoy them when they taste good. The ones we find that work the best are hard root vegetables, which we cut up into large chunks and cover with the slurry in a jar that we allow to ferment at room temperature. Beets, carrots, parsnips, celeriac, and sunchokes are among our favorites. We feel that they're optimal after somewhere between three to five days, when there's a nice balance between the sweet and the tart. You always have the option of cutting the root veg into smaller pieces to make the infusion happen faster. However, it'll likely be a sweeter pickle. When it comes to other vegetables that aren't as hearty, it only takes somewhere between a couple of hours to overnight. Sliced cucumbers will become mush overnight. Sometimes all you need is a light coating to do the job. Think of anything that isn't a root vegetable as a quick bread-and-butter pickle. Split whole spring onions with the tops intact are wonderful.

We've found that you can adapt the original bettarazuke technique of using whole daikon with other root vegetables that have the same type of structure. When it comes to salt rolling and pressing to achieve the desired effect, we separate vegetable chunks by size into various batches, and can

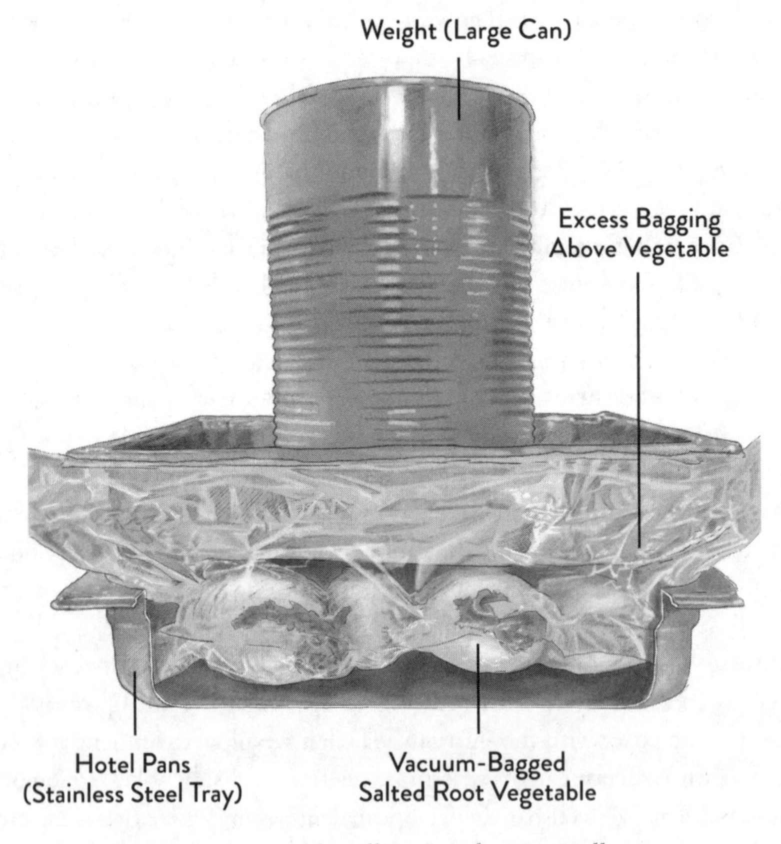


Setup for pressing salted vegetables. *Illustration by Max Hull.*

use less salt by bagging them in a layer using a Cryovac or FoodSaver. If you don't have either, plastic zip-top bags can work. In general, this method works best with long root vegetables shaped similarly to daikon, such as carrots and parsnips. As a guideline, you can work with anything that doesn't exceed the diameter of an average daikon. When you choose vegetables, find those of similar diameter, and of the shortest height when laid down. Locate pans that will nest into each other (hotel pans work great for this). Map out how the vegetables will create a single layer in the pan when bagged. To maintain the layout in the bag, start with the vegetable that will go inside first. Pour a healthy layer of salt that's roughly twice the size of the root on a large cutting board or clean surface. Place the root vegetable at the middle of the pile, then apply pressure while rolling until it starts to weep liquid and the crystals adhere. Coat with salt and place it in the bag.

Continue for the remaining vegetables as they're organized in the bag. Vacuum-seal the bag. Place another tray on top and put a heavy weight on that. If you don't have a weight, tie-down straps work well, but you'll have to increase the tension as required. The next day, put a tiny hole at the corner of the bag farthest from the vegetables, mark it, and keep it as high as possible to prevent leakage. This will allow the carbon dioxide from fermentation to escape. Once the vegetables are done compressing after a couple of days, remove, rinse, and add to the kojizuke brine.

## Beet Koji Kvass

Overwintered root vegetables tend to be more starchy than sweet. Therefore it makes sense to brighten up a beet kvass with an ounce (30 g) of koji per pint (500 ml). Kvass is typically made by filling a jar with large chunks of cut-up beets, sprinkling in a bit of salt, covering them with water, and allowing them to lacto-ferment. It's a satisfying savory elixir that is often used as a flavor driver for borscht. Sometimes starter cultures are used to get the process going, as with pretty much any lacto-ferment happy product such as sauerkraut or yogurt. That being said, leveraging koji as a starter culture is a great fit. Also consider that the enzymatic activity for a touch of sweetness and umami makes the kvass quite nice.

## ARTISANS HONORING TRADITION

### Kevin Farley

*The Cultured Pickle Shop in Berkeley, California, is an artisan fermented food and drink company that we've admired and respected for years. Owners Alex Hozven and Kevin Farley have created some of the most delicious products by following traditional fermentation practices, and as far as we know, their shop is the only place outside of Japan where you can get* tsukemono *(Japanese pickles) at a craftsmanship level. Their focus on flavor, attention to detail, and use of local organic products is unparalleled. Their pickles are not only enjoyed at kitchen tables across the Berkeley area, but also sourced by fine dining restaurants across the US. What they are doing is really special. We are honored to share how they got started, a day in the life, and the importance of sharing. There's also a deep dive into how to make kasuzuke, an amazing pickle made from the lees/dregs from sake making. We couldn't find a better voice for their journey and methodology than the gentleman who lives it every day. We hope Kevin inspires you to seek out more products from folks who truly care about what we eat.*

Credit for the existence of the Cultured Pickle Shop goes to Alex Hozven. For over twenty-five years, she has deftly applied her tremendous talent and tenacity to growing a business that not only supports our family but has taken an ancient tradition of food preservation to a level of craft that has proven her to be well ahead of her time. Years before the charismatic and elegantly mustachioed Sandor Katz crisscrossed the country extolling and proselytizing for a fermented revolution—leaving in his wake bubbling and gurgling jars from Williamsburg, Brooklyn, to the Mission District in San Francisco—Alex was putting in eighteen-hour days, day after day, month after month, year after year. She can be ascetic and hermetic in her work, spending long hours alone in the cave overseeing and guiding nearly two hundred independent ecosystems. To this day she doesn't approach her work as a job but as an art, participating and crafting at every level of production. Each jar of sauerkraut, each clove of miso-pickled garlic, each bottle of kombucha has come from her hands and her heart. If you've ever tasted them, you know. I believe that her approach to craft in commercial production is what *artisan* truly means. It has been the great honor and pleasure of my life to work alongside her.

On an average day we begin work in the predawn hours. We make tea for that day's kombucha. We tend to cave duties together. We inspect the tanks for cleanliness and to make sure the lids are secure. We monitor the ferments to be sure that they are progressing appropriately. Much of my early morning is spent in intimate proximity with the ferments. Those tanks in their first week and in their fourth week are dismantled and I push down the kraut to expel $CO_2$, monitor the viscosity of the brine, and taste. We prep the kitchen for that day's work. I run home around 7 AM and wake the boys up, fix breakfast, pack lunches, and get them off to school, then it's back to The Shop for a day of prepping vegetables, filling jars, or delivering product. On Saturdays, while Alex is at the Berkeley Farmers' Market, I work on small-scale experiments to better our understanding of certain methods we have been using for years and to see if there may be some new avenues to explore in the coming years. Many of these experiments wind up posted on social media, but I have yearned for a better format to communicate and document our work and to tell our story.

After many years with our heads down, alone in the cave with the bubbling tanks, I have been forging relationships with chefs and other food-minded people around the world. These connections have lead to conversations, questions, and exchange of information that have made us even better at what we do. That is all we have ever wanted. We have never been interested in getting the product into as many hands as possible. We have never really been interested in running a business. We haven't seen ourselves as part of a movement, or as educators or proselytizers for a better way of living. We have seen ourselves as artisan producers of artisan products, steeped in this strange obsession with process and outcome. So much of what's being said or written about fermentation these days revolves around health or personal food security, and though these conversations are incredibly important, increasingly so in our modern culture, after twenty-two years I find them tiresome. These qualities are inherent in what we do, and far more intelligent people than I have already done a great job of articulating them. So I think here I will reach for nothing more lofty than illuminating and exploring our pathways toward deliciousness. Which for me is quite lofty enough.

### KASU, SAKE KASU, LEES, DREGS

After a tank of sake has run the course of its fermentation—anywhere from eighteen to thirty-two days—what remains is a white mixture of sake, rice solids, and yeast. This mixture, known as the moromi, is pressed to separate the sake from the suspended solids. There are several methods of pressing the sake out, leaving compressed rice solids, or lees, behind. Larger sake producers extract the sake from the lees, called kasu, by machine, and the solids that come out in thin dry sheets called *itakasu*. Smaller producers often press their sake by hand using a wooden box called a *fune*, which has a lid that gets cranked down on the sake mash, or moromi, which has been placed in small canvas bags. This method will yield a kasu that is moist and chunky called *teshibon* or *namakasu*.

Kasuzuke, or vegetables pickled in kasu, are said to have originated in the Kansai region of Japan as early as twelve hundred years ago. The first vegetables known to be fermented in kasu were white melon, named *shiru-kasu-zuke* or *Narazuke*. Later, the technique would be used with cucumbers, eggplants, and uri (bitter melon). It was produced primarily by Buddhist monks and used by Samurai for sustenance in wartime and winter. During the Edo period of the seventeenth century, sake producers promoted the use of kasu widely. Not long after, kasuzuke would become a mainstay in the ever-expanding repertoire of Japanese tsukemono.

We are fortunate to be located just blocks from Takara Sake, one of the largest producers of sake in the United States. We have a long-standing relationship with Takara, and after each pressing of their certified organic Ginjo-grade junmai nama sake, they put aside about 150 pounds (68 kg) of kasu for us. The kasu has been pressed nearly dry at the factory and has a slightly sticky, putty-like texture. Kasu is stored in the refrigerator; the cold makes it stiff and a little unyielding. We allow it temper in a bowl on the counter for the day, and as it warms it becomes much more pliable and easy to work with.

We begin with a two-day press of the whole vegetable with 6 percent salt. The purpose of this is to move some liquid out of the vegetable before it goes into the vessel, which lessens how much it will crash out in a salty medium and helps with texture preservation. We then build the kasu base mixture with a ratio of 10 parts kasu to 3 parts sugar to 1 part salt, with a yearlong fermentation in mind. This works well for root vegetables such as burdock, beets, and sunchokes. We also regularly use green garlic, spring onions, jalapeños, and kabocha squash. In a stainless steel vessel, we layer the kasu and vegetables, cover, and put into our fermentation room, which averages about 65°F (18°C), to age. For the first six months or so the results are fairly unpleasant—the salt is harsh and forward on the palate, the sugar is cloying, and as it's metabolized the alcohol can be intolerable. At around nine months we start to check for doneness.

We are hoping to observe and capture a moment of balance within the vessel where a harmony is struck between preservation and transformation. For us, usually between the ninth and eighteenth month, we look for a moment when the salt has softened and rounded out, when the pickle is sweet but not cloying. We want the essential vegetable qualities, both flavor and texture, to remain. There are elements of a long braise in this pickle, a slow caramelization that builds over time. When we sell our kasuzuke, the kasu itself has gone from white to tawny to dark brown. There is a moment as you bite that the vegetable begins to melt as deep, rich—low and slow—salts and sugars begin to flood your mouth, and then a sudden snap and crunch as the flavor of fresh vegetable fills the rest of your bite. It's a delightful pickle. If left to continue we see those caramel flavors and colors move from light to dark. As we reach the fifth year, we see flavors range through malt and chocolate. As we move past ten years, we reach tobacco, blackberry, and licorice. Of course by this time only a slight texture difference may distinguish the vegetable from the kasu.

When we package this pickle, we jar up the vegetables with a bit of the kasu so they continue to ferment at home. The remaining kasu we keep and treasure. It is a wonderful product, preseasoned and full of flavor. We use it to season soups and stews, we make dressings and sauces with it, and we use it to ferment tender vegetables such as leafy greens or mushrooms and animal proteins such as black cod and pork shoulder for hours or days.

### BURDOCK KASUZUKE

Burdock. *Arctium lappa. Gobo* in Japan. It is the long taproot of a thistle. Dark and woody, it is deep in earthy overtones and slightly sweet. We wash the burdock well and press it at 6 percent salt for two days. The burdock is spiraled into a vessel and layered between the kasu-sugar-salt mixture. In the initial weeks of fermentation, there will be a fair amount of carbon dioxide released and pockets will appear. To combat this, and ensure optimal kasu-to-burdock contact, we weigh the ferment down.

Our burdock ferments for twelve to eighteen months. One of our longer ferments, it needs that duration for the dense, sturdy root to ferment clean through. It is a testament to the hardy strength of burdock that after a full fermentation it retains most of its flavor and texture. Though the sake and koji permeate, it is still astonishingly earthy and woody. For years we enjoyed this pickle sliced thinly on a bias. Wood chips, I called them. Wonderful, flavorful wood chips. Then I introduced it to the Microplane. Burdock kasuzuke loves the Microplane. A melt on your tongue, snow of sweet earth.

## KINZANJI MISO

### Jeremy Kean

*Jeremy Kean is the chef-owner of Brassica Kitchen + Cafe in Jamaica Plain, Massachusetts. He co-owns the restaurant with his sister Rebecca, and his best friend Phil Kruta, putting on some of the most interesting French-technique-inspired food in the Boston area. We first met a few years ago; Rich still remembers the first time Jeremy K. tasted a yogurt miso hot sauce and was mind-blown. At that moment, he asked that we make him 20 pounds of koji so he could make amino pastes and sauces with everything and anything he could think of. Since then, he's gone through all those ferments and has adventured out on his own. One make in particular that we're excited about when it comes to utilization of a variety of products at once is* kinzanji *miso. Here is his story on the spinning of a traditional method.*

I first learned about kinzanji miso from a Japanese miso maker who posted about it on the internet. My interest was driven by the seasonal influx of vegetables from time to time at Brassica. While I was recently in Japan, I inquired about the uses of this special miso and learned on two accounts that it is used as the French use mirepoix and Italians use *soffritto*. I was fascinated. Traditionally, the koji is made with barley and soybeans; then a variety of mixed, minced vegetables are added as the base for the miso.[1] Based on this idea, the koji we used was half jasmine rice and half barley. The vegetable portion consisted of anything and everything we had in house: diced radish, sunchoke, almonds, macadamia, Cara Cara orange, and lacto-fermented garlic. I also decided to add toasted rye berries along with cooked soybeans and barley. Everything was mixed together with salt and pressed into a giant crock. After two weeks, it was very interesting (although not mind blowing) in flavor as it had not undergone the traditional six-month fermentation period. However, I had other plans for it. To me, it had all of the flavor notes of a caponata: sweet, sour, salty, savory and a touch of bitterness, but not the depth and marriage. It was pretty much all the driving components of the dish prior to cooking it. The application of heat and melding of all ingredients is what made it unforgettable.

**1 part koji**
**1 part minced vegetables (anything you have that matches, based on past cooking experience or not)**
**4 percent salt**

Mix all the ingredients together and contain as you would any other amino paste. Hold for 2 weeks at ambient temperatures, then use for the caponata recipe below. If you think of this as a beginning for any delicious sauce that starts with the aromatics of mirepoix, soffritto, or any other aromatic vegetable base used all over the world, the possibilities are endless.

## KINZANJI CAPONATA

This caponata is delicious and the perfect way to use the kinzanji miso in the recipe above.

**1 cup (250 ml) good olive oil**
**5 shallots, sliced**
**1 cup (200 g) sugar**
**4 cups (1 kg) kinzanji miso**
**1 bottle (750 ml) cava**
**0.25 ounce (7 g) fresh-cracked pepper**
**1.75 pounds (800 g) breaded and fried eggplant (fried till almost, but not quite, burnt)**

Heat the olive oil in large pot and sauté the shallots with the sugar until a caramel is formed. Add the miso and stir constantly until it looks like a thick tomato sauce, but not quite a paste. Pour in the cava, sprinkle with pepper, and reduce until it's the same thickness as before. Allow it to cool. While you're waiting, this is a good opportunity to fry the eggplant if you haven't already. Rough-chop the fried eggplant and mix everything together. One of our favorite ways to enjoy this is spreading it on slices of grilled, crusty bread.

## TAKING USE-EVERYTHING SERIOUSLY

**Kim Wejendorp**

*Kim Wejendorp is the head of research and development at Amass Restaurant in Copenhagen and a good friend of ours. We have been following his work and exchanging ideas for a long time, and are always inspired by the breadth and depth of his investigations. When it comes to questions about the science behind fermentation, he's at the top of the list of chefs we turn to. One thing that we find amazing is Amass's undying motivation to use every last bit of all products. Matt Orlando, the chef-owner, has this philosophy: "No such thing as a by-product, only another product."*

I encountered koji my whole life without realizing it. Indeed, most people are surrounded by it on a daily basis. Soy sauce and miso are quite common these days, and very few people really question how these products are made. It wasn't until I moved to Japan that I began to take an interest in the process. Every year during the autumn season, the residents of the town where I lived would meet together to make their own miso from the beans they had grown in their gardens. The knowledge of the making being passed on from grandparents to the younger generation was ingrained. It was here I learned the basics not just of mixing ingredients but also of the magic that could occur. Tasting a quick shio koji pickle, last year's miso, a homemade soy sauce, then a treasured ten-year-old miso . . . so much deep flavor. It was here my interest in *Aspergillus oryzae* and the incredible work (and variety of jobs) it can do began.

These days I still make my own miso, but I also use the mold in combinations that I never would have thought of back then, especially in Japan. At Amass we are constantly striving to serve delicious food while being responsible by using as much of the ingredient as possible. This means finding ways to use not only the lesser-thought-of cuts of meat or types of fish but to use the peels, cores, seeds, guts, and so on.

The approach is not only to serve them as a novelty but to make them delicious by treating them the right way. With enough thought, these things can be turned from food waste or by-products into something edible—products with their own intrinsic value.

Today *Aspergillus oryzae* is a tool we use to make a wide variety of products with what most kitchens compost or discard. It's used to break down the proteins left behind in the nut pulp from making almond milk. We also turn our excess egg parts and fish trim into rich seasonings and condiments. We convert the starches from old bread into sugar to feed a vinegar, and vegetable scraps into deeply rich cooking (or drinking) alcohols such as mirin.

## CARROT SCRAP MIRIN

Mirin originally was a rice wine enriched with inoculated glutinous rice. The advent of distillation brought stability to the process and set it on the road to the modern condiment we know today. It was popular during the Muromachi period (1336–1573 CE) of Japan as a luxury drinking liquor. Mostly used as a cooking rice wine, a well-made hon mirin can also be a pleasure to drink akin to drinking a sweet wine or sherry. In mirin a high percent of alcohol is used to preserve inoculated grains, allowing the saccharification process to happen without fermentation. During the booze bath the grains give up their sugars and their proteins split into basic aminos, adding a gentle, savory richness. Here's how you can make a delicious mirin from carrot scraps.

**4.25 cups (1 L) 151-proof neutral spirit**
**4.25 cups (1 L) carrot juice from scraps**
**2.2 pounds (1 kg) rice koji**

In a large bowl, mix the spirit, carrot juice, and crumbled koji together. Pour into a mason jar and leave in a cool, dark place for 6 months to a year or until you can't wait longer. Strain out the solids and enjoy as an aperitif.

# No-Waste Citrus Miso

In the interest of the waste-nothing approach discussed by Kim Wejendorp (see the sidebar "Take Use-Everything Seriously" on page 240), we've been talking to him about a crazy make he once told us about, a miso he made from what remains after juicing lemons. We found the concept hard to believe until he pointed us to literature on the debittering enzymes koji generates as it grows.[2] What convinced us of how well it worked was a care package that allowed us to taste it. The flavor is best described as a lemony olive with a tad of bitterness. It opened our eyes and made us think about the further potential of this application.

**3 kg pearled barley**
**1 kg barley koji**
**1 kg juiced lemon skins**
**300 g salt**

Cook the pearled barley and allow it to cool. Mince all of the ingredients together, then purée in a food processor. Pack into your chosen miso container and press down with a weight. Ferment at room temperature for at least 6 months.

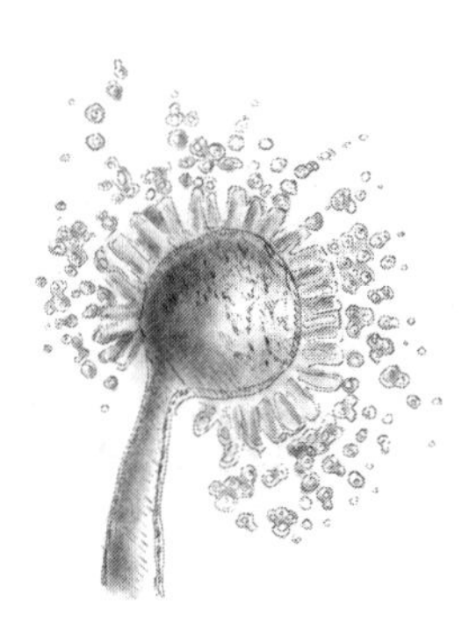

CHAPTER 13

# Sweet Applications and Baking

Traditionally, koji was primarily used to power secondary applications and not typically utilized as a fresh ingredient. However, its inherent sweetness and touch of umami make it a fascinating sugar base to work with. Still, using fresh koji immediately for its base characteristics is a tough sell if you're not producing at a relatively high level. Almost everything we've discussed in the book so far has been about leveraging the power of enzymes for investment in flavor, and preservation that requires waiting at least a week and extends to a year or more. That being said, we've found value in showcasing sweet koji as a driving ingredient. We think it's worth your time, and you'll understand once you have a ton of it made waiting for a purpose.

Throughout this book we've shared ideas of optimizing conditions to maximize sugar and amino acid production with koji enzymes. Not long ago the key to good eating—aside from basic nutrition—was using just enough sweetness to make something taste good, which is really a matter of finesse. Before refined sugar was readily available, folks depended primarily upon seasonal fruit and honey to include sweet treats in their diets. Today candy, soda, sweet desserts, and treats surround us everywhere we go, and this level of sweetness is what people expect when you tell them something is sweet or has sugar in it. What we're talking about here, however, is how to add a *touch of sweetness* so your diners can experience the subtleties of the base flavor.

The beautiful thing about long-grain rice koji grown with a sugar-focused *Aspergillus oryzae* is that it gets quite sweet at the forty-eight-hour mark. It's especially sweet if you run your incubator slightly hotter at 95°F (35°C) to encourage the generation of amylase enzymes. However, it's still not as sweet as a natural syrup like honey. We find its sugar level and starch combination comparable to a ripe pear. To us, that means it's wonderful but difficult to maintain the nuances of flavor without overwhelming it in an application.

Koji's ability to add a touch of umami to make something delicious is especially important here. Think of all the savory crossovers that came into existence that were once considered strange but are now fairly accepted: tomato jam, bacon ice cream, miso caramel sauce. Koji made from protein-rich grains provides just enough to give flavor depth to sweet ingredients to bring everything together and make it wonderful to eat, changing your dessert game.

## Koji for a Touch of Sweetness

Two forms of koji work well for immediate use: fresh as is, and *amakoji* (a thicker version of amazake). The fresh applications are just that, using koji as an ingredient without leveraging the enzymes or fermentation. Amakoji is used when the amylase has converted starch into sugar to make it sweet but hasn't created notable acidity or even alcohol. To make it, all you need is 1 part koji, 1 part cooked grain, and 1 part water. Blend and let it sit out at ambient temperatures until it's sweet and hasn't soured. A couple of days is usually enough to get you there. You can also just use the amazake we detail earlier in the book.

## KOJI-BASED TREATS

### Allie La Valle

*Allie La Valle, Jeremy's wife and chef-owner of Larder, told us that nearly everything from amino pastes to shio koji to fresh koji itself can be used for desserts; it really boils down to using not just what you have on hand but also what you enjoy. She has given us some suggestions to pass along to aid you in your own exploration:*

- Caramel loves amino pastes. Simply mix in a touch of amino paste as your caramel cools. A delicious example of this is to make a caramel apple pie filling. This is especially delicious if you work a hard cheese such as Parmesan into the crust due to how well the umami of these cheeses marries with that of the amino paste.
- Steep fresh koji in milk or cream as you are cooking it to make a custard or pudding. The fruity and floral notes of the koji pair amazingly well.
- Add a splash of an amino sauce to your brownie recipe to reinforce the deep earthy fruitiness of the chocolate.
- Dried and toasted ground koji rice goes great with anything containing coconut. It is really nice when added to coconut macaroons.
- Fresh black koji puréed with sugar and then spun in an ice cream maker creates a delicious frozen dessert. Use about 0.5 cup (100 g) sugar per pound (450 g) of koji. Adding black trumpet mushrooms takes it to a whole other level.
- Grow koji on oats and use it to make oatmeal cookies. Plump some raisins in amazake and add them for a real treat.
- Puréed and strained amazake that has been cooked until thickened serves as an amazing glacé for fresh fruit tarts or as a glaze for puff pastry and Danish dough.

One of the techniques that we are absolutely in love with is adding pieces of fresh fruit to koji. We're talking about fruit that still needs time to ripen, like the stuff you typically see at the grocery store. It's a pretty perfect application—leveraging inherent starches to create sweetness as well as softening the structure. You might say that it's a fruit ripening cheat. No matter what you call it, it makes fruit taste great. By no means is it the same as a beautiful naturally ripened product, but it is quite nice.

We've found that koji works well for apples, pears, stone fruit, muskmelons . . . pretty much whatever falls into a similar flesh-structure category. Softer fruits tend to break down easier, so it's easy for them to get mushy. (When you're at that point, a purée for a sorbet is a wonderful application.) To prepare koji-infused fruit, we typically cut it into large pieces to be sliced nicely after the fruit has been macerated. In terms of size, it all depends upon how much infusion you want. For a muskmelon, it ranges from cutting it into eight wedges with the skin intact to 1-inch (2.5 cm) cubes. It all depends upon the level of infusion you're looking for in the application. We suggest testing a variety of shapes, sizes, and durations to find what works best for specific fruits. This approach is no different from marinating pieces of protein.

There are two ways to apply koji to fresh fruit. The first is making an amakoji slurry to bury or coat the fruit in. All you have to do is blend the amakoji you made until it's similar to the consistency of a loose polenta or porridge. Put cut fruit in a bowl or container, then add enough of the amakoji so the fruit is coated when mixed. Cover the bowl and allow the fruit to macerate at room temperature overnight (or less time, if you're satisfied with the flavor). Wipe off the excess if so desired, slice, and serve.

The second method is vacuum-packing straight-up fresh rice koji with something as simple as a FoodSaver or, even better, a commercial Cryovac bagging system (for those who have access to one). You can make a paste with the fresh koji and a touch of water to make it easier to coat. Mix cut fruit with the koji mash until all exposed surfaces are covered. As with all bagging, invert the last couple of inches to prevent messing up the sealing area. Set the bag on a flat work surface and place the coated fruit in an orderly fashion so they don't overlap one another when the vacuum is applied. Fold the back back to its original state then vacuum-seal. We've found that the infusion of the koji sugar makes the fruit taste great if you open it immediately after you've sealed it.

# Hot Koji Kombucha Arnold Palmer

In the colder months of the fall and into the winter, we like to provide a comforting drink to folks who come to our workshops. We also enjoy putting spins on popular ideas we know and love. One of Rich's favorite drinks is an Arnold Palmer, also known as a "half and half," which consists of equal parts lemonade and tea. It's a wonderful marriage of sweet tartness and earthy depth. Understanding these flavor components, we often change the ingredients up based on what's in season and available. Recalling how amazake was traditionally served hot in the winter, it only made sense to use it as a sweetener for a drink to be served at a koji class. Rich often enjoys the challenge of replacing the acidity of citrus with homemade vinegars to make cocktails. So why not kombucha in an Arnold Palmer? The flavor profile is already pretty close. However, just mixing amazake and kombucha won't quite cut it. Unless your kombucha is young, its tartness will be too overwhelming for a proper mix to reach a drinkable consistency with the amazake. The answer is to dilute it with a freshly brewed tea to create a well-balanced beverage worthy of serving as a winter warmer. It also doesn't hurt to add some cider.

Mix 1 part kombucha, 2 parts well-blended amakoji (alternatively, 1 part sweet rice koji and 1 part water, blended), 2 parts tea, and 2 parts apple cider. Heat to a simmer and serve.

# Hot Miso Milk

When we are talking about sweet enhancements, we're not just talking about leveraging koji for the starch-to-sugar conversion. We are also interested in how the complexity of the long-term koji makes, such as amino sauces and pastes, enhances sweet drinks. Whenever you think about the quintessential drink of your childhood to warm up, isn't it hot chocolate? That rich, creamy, and sweet delight that you'd often risk burning your tongue to enjoy. Hot chocolate has an indescribable, intoxicating flavor complexity and depth that draws us to it. Interestingly enough, miso has the same qualities even though the flavor profiles are very different. This got us thinking that this would be a wonderful replacement to make a hot drink with.

The beautiful thing about an aged dark miso / amino sauce is all you have to do is add a touch of sugar and some water and you end up with this crazy umami caramel-like sauce. If you think about how the Maillard reaction occurs over time to make miso dark, it only makes sense that, when sweetened, it would taste somewhat similar to caramelized sugar. However, the umami hit gives it a distinct advantage.

The easiest way to make hot miso milk is to make miso sugar. All you have to do mix 1 tablespoon of dark miso to every 1 cup (200 g) of granulated sugar that you plan to make. It's not an issue if you make too much, as it can be stored like brown sugar. Bring the milk to a simmer and stir in the miso sugar to your preferred level of sweetness. If you don't quite get enough miso, add a touch until the complexity comes through. As with any recipe that we walk through, milk and sugar can be in whatever form you have or that fits in the same category of flavor and texture. This application works great with all the non-dairy "milks." (Sidenote: If you're loving the current oat milk craze, you can make oat milk amazake simply by making amazake with oats instead of rice—it's way better!) Sometimes it's just a matter of balancing with water as needed. In the same vein, explore other sugars.

## Horchata

Amakoji can be used as the base for horchata, a delicious sweet, creamy, and spiced drink made from rice milk or tigernuts, which originated in Spain and spread to much of the Spanish-speaking world. For the amakoji base, we recommend a long-grain rice koji that has been grown with an *Aspergillus oryzae* for sake to maximize the sweetness. Making the horchata is quite simple. Dilute the blended amakoji by mixing enough water to bring it to a drinking consistency. It's around 2 parts amakoji to 1 part water. Once you're happy with how it drinks, taste it with a pinch of cinnamon mixed in. You'll find that it's refreshing and filling at the same time. If you find that it's not quite sweet enough for your liking, feel free to add a touch of sugar. We find that maple syrup works quite well. Once you get it where you like it, serve in cups and dust with a little cinnamon on top.

## Larder Sodas

At Larder, Jeremy makes fresh seasonal sodas to order by blending syrups with seltzer. These sodas range from chocolate phosphates and egg creams, to foraged-root root beer, to unique ones like chanterelle and, of no surprise, koji. The koji sodas rotate with the seasons—say, koji melon or koji peach in the summer, and koji pumpkin in the fall and winter. Making these sodas is simple and can easily be done at home or at your bar. In the case of the koji melon soda, a base is made by combining 1 part cantaloupe and 1 part amazake and blending it smooth. Sometimes an addition of 1 part sugar is needed, as this base needs to be sweet enough to hold up to dilution by the seltzer water. This ratio works with any ingredients you think will make a delicious drink, so experiment. Simply add 1 fluid ounce (30 ml) to 8 fluid ounces (237 ml) of seltzer and stir to mix. The addition of alcohol, especially a koji-based one, is never a bad idea.

## Rice Pudding

Amazake by itself becomes a delicious rice pudding when cooked down. Just low-simmer it in a pan on the stove until it reaches the consistency you desire. If you're not feeling quite as adventurous, feel free to use 4 parts amazake, 4 parts cooked rice, and 1 to 2 parts water and follow the same process. Add a sweetener and finish with minced dried fruits and vanilla or cinnamon as you desire.

# Koji Champurrado

Rich recently became enamored with Mexican ingredients thanks to his friend James Wayman. They collaborated on a koji-driven dinner that featured dishes inspired by James's travels to Oaxaca and Tabasco. Aside from creating an insane depth of flavor in the savory dishes, they wanted to feature what koji had to offer on the sweet side of things. Rich did some research on drinks and happened to find *champurrado*, a comforting chocolate-and-masa-based hot drink that's similar to hot chocolate. It's sweetened with *panela* (unrefined sugar) and typically flavored with cinnamon and vanilla. James had made this amazing masa koji with Oaxacan corn that wasn't sweet, but the transformation elevated the flavor in an indescribable way. It only made sense to use it in the drink. For the sweetener, Rich used amakoji made from jasmine rice along with just enough panela for a touch of sweetness. He used the chocolate from an organic farm in Tabasco and allspice straight from a famous Sunday market in Oaxaca. The complexity and depth of the Mexican ingredients elevated the drink to another level. When Rich tested the full composition of the drink for the first time, James told him it was one of the best things he ever drank. The basic version shared here is a delicious alternative to hot chocolate with accessible ingredients.

**1.33 cups (200 g) nixtamalized corn koji**
**1 cup (200 g) amakoji**
**2.33 cups (550 g) water**
**0.75 ounce (20 g) Mexican or quality dark chocolate**
**2–4 ounces (50–100 g) panela or dark brown sugar**

Purée all the ingredients, except for the sugar, until smooth. The consistency should be a little thicker than you're used to drinking, but not as thick as a porridge. You may need to add a little water, as corn koji hydration varies. Once you get it to the proper consistency, add it to a medium pan. Heat over a medium-high flame while whisking occasionally until it comes to a simmer. Cut the heat to low and whisk in 2 ounces (50 g) of sugar. This is the amount that we recommend—it allows you to taste all of the elements without the sugar overwhelming. Please do not expect it to taste as sweet as a hot chocolate. Feel free to add more sugar to your liking. The recommended serving size is 3 ounces (90 ml), as it is a rich drink. A really light touch of both ground allspice and finely grated dark chocolate on top for aromatics just before serving is highly recommended.

## Puffed Koji

Puffed grains and seeds are enjoyed the world over, from popcorn to Rice Krispies. Nearly everyone likes them. Here's our technique for puffing fresh koji and a way to make puffed koji rice treats.

**0.66 cups (100 g) fresh koji**
**2 quarts (2 L) oil**

Preheat your oven to 170°F (76°C). Spread the rice out in a thin layer on a baking sheet. Bake in the oven for 3 hours. The rice should be very dry and hard when it's done. Allow the rice to cool for at least an hour.

Heat the oil to 350°F (176°C). Fry the rice in small batches for about 30 seconds or just until the rice puffs up. This happens very quickly and you can easily burn the sugars in the koji so pay close attention. Use a slotted spoon to remove the rice and place it onto a paper towels to drain.

We like using this rice to make crispy rice treats. The ooey-gooey marshmallow pairs great with koji. This puffed rice also makes a great substitute for oats if you're making granola and can be mixed with miso powder, nori, and toasted sesame to make a great topping to sprinkle on about any food that you want to add a savory crunch to.

## ANNA'S KOJI SORBET

### Anna Markow

*Anna Markow is a pastry chef in New York City with a wealth of understanding of techniques and methods as a result of years of dedication to the craft. She has an unbelievable palate that allows her to immediately parse out half a dozen different applications upon tasting something completely new. One thing that we've always been impressed by is her ability to marry flavors, especially if they're not traditionally used in a dessert or pastry. Knowing what she can do, we shared sweet koji with her to see what she would come up with.*

Koji is extremely versatile for pastry. The floral aroma and natural sweetness from a batch of jasmine rice koji gets creativity flowing toward complementary ingredients like berries and stone fruit, and, while using the koji to make ferments and amazake were also on the docket, capturing the flavor of the koji itself was something I knew I had to do to test its full potential.

In order to taste the true essence of koji, I created a recipe for koji sorbet, which, thanks to the starch in the rice, churns up the texture of a fine gelato. To achieve the right consistency and ensure the sorbet would stay soft enough to scoop, I found the Brix level and compared it with recipes for pulpy fruit sorbets with a similar sugar content, then adjusted for flavor as I blended, adding a small amount of water to ensure smoothness.

The sorbet is extremely sweet with a cultured tang, making it a surprising alternative to frozen yogurt, and a perfect foil for whatever accompaniments, fruit or otherwise, you might want to use. Anything from fresh sliced strawberries to tangy fermented blueberry pie to salty miso caramel sauce would play nicely with this sorbet. It also makes a very bracingly refreshing probiotic float when added to a glass of kombucha.

**500 g jasmine rice koji, fresh or frozen**
**50 g granulated sugar**
**Up to 200 g water**
**Salt, to taste**

Place the koji, sugar, and a splash of water in a blender and begin puréeing on medium speed. Drizzle water into the blender slowly to help loosen the purée, then turn the speed up to high and continue adding water to make a thick but perfectly smooth and fluid paste. Blend in salt to taste.

If blending has warmed the koji purée, chill it thoroughly before processing in an ice cream maker according to the manufacturer's instructions. Remove to a plastic quart (liter) container and pack tightly, pressing plastic wrap over the surface before freezing completely.

## Using Koji in Baking

There are no boundaries when it comes to baking with koji. Adding the smallest touch of umami changes everything; from breads to pastries, there are numerous ways that koji can be incorporated. Over the years we have amassed a plethora of uses for koji that center on baked goods, all of which are simple and straightforward and with minimal tweaking can be easily incorporated into your existing recipes. The great thing is that most of the koji-derived ingredients that can be added to an existing recipe have already been covered in this book. From here it is just a matter of using them within an existing framework. Koji-cultured butter (see the "Amazake-Cultured Cream" recipe on page 215) is a prime example of this. You can substitute this for the regular butter that you would normally use in your existing recipe for shortbread cookies or caramel or brownies or pound cake and the list can go on and on.

Our journey using koji in baked goods started with a desire to add extra depth to these foods. While we all enjoy a sweet treat, sometimes they can be too sweet, essentially one-note foods with little to no complexity.

Ask virtually any pastry chef about this and they will stress the importance of balancing various tastes within any given dessert or baked good. A bit of salt, a splash of sour, and a pinch of umami will elevate something beyond a sweet treat into intensely memorable. It should be noted that while cooking can be compared to acid jazz loaded with spontaneous improvisation, baking is more akin to a strictly composed and conducted orchestra. This realization that the science of baking is just as important as the art and craft led us to find ways that koji could bring interesting harmonies to the already existing constructs rather than try to reinvent the wheel.

With basic experience and understanding of the principles of baking, you can easily see many windows of opportunity that can be opened. Take the "Sour Amazake" recipe on page 115, for example. It contains far less rice and koji solids than sweet amazake and has a pleasant sour taste akin to sourdough breads. By straining this and using the liquid to replace some or all of the water in a bread recipe, you can create koji breads that are unique and delicious.

One of the most impactful ways to introduce the joys of koji to a baked good is via an amino paste. This immediately adds a depth of flavor that cannot be achieved otherwise. The beauty of using an amino paste is that you don't need much and, for practically any recipe, no adjustments need to be made. For context, think about how salt sprinkled on top of chocolate chip cookies makes the flavors pop as you chew. Now consider what happens when you add an amino paste that already has flavor complexity way beyond salt or even the caramelization of sugar. The dimensions of an aged miso are on the level of chocolate and coffee, but the kicker is that it has umami.

For perspective, let's talk about one thing that will help with understanding: the difference between brownies and blondies. At the recipe core, the only difference is cocoa versus brown sugar. However, this creates such a drastic change in the complexity of flavor that you can't really put the two in the same category. Also, consider that blondies commonly have chocolate chips in them. Hmm . . . Let's say that adding complex flavor components to something to make it shine really isn't a stretch. You may even need to make miso blondies.

# Allie's Tahini Cookie

Jeremy's wife, Allie, has developed a tahini cookie at Larder that pairs fantastically with an amino paste made from great northern beans and *Aspergillus sojae*. The amino paste adds a layer of fruity umami to the cookies that elevates them from delicious to divine. In discussing Allie's inspiration behind this, she told us that she found many flavor and aroma similarities between this amino paste and tahini. In experimenting with how much of this amino paste to add, she simply replaced the salt from the original recipe with amino paste. Allie elaborated that it's as simple as tasting the batter to see if it's delicious.

**9 g amino paste, short-term preferred**
**139 g butter, softened**
**139 g sugar**
**68 g honey**
**154 g tahini**
**267 g all-purpose flour**
**4.8 g baking powder**

*Note*: For baking applications, we recommend using grams to measure rather than the US equivalent. The precision is key to success.

Preheat your oven to 350°F (176°C). Cream the amino paste, butter, sugar, honey, and tahini on medium speed until fluffy, about 3 minutes. Mix and sift the flour and baking powder together. Add the flour mixture to the creamed butter mixture and mix on low speed for 3 more minutes or until the dough is smooth. Use a 2-ounce scoop to portion the cookies onto a parchment-lined baking sheet. Bake the cookies for 15 to 18 minutes until they are golden brown around the edges. Allow them to completely cool before removing from the baking sheet.

Yields 15 cookies

# Chef Markow's Snickerdoodles

Chef Anna Markow uses koji not only in desserts such as sorbet (see the "Anna's Koji Sorbet" sidebar on page 252) but also in baked goods such as snickerdoodle cookies. Koji, however, poses a tougher challenge with these, as its aggressive sweetness can threaten to overshadow other elements on the plate. The high sugar content makes caramelization a logical next step, but evenly toasting the grains is tricky. Low and slow is the key, resulting in relatively consistently roasted grains with a fully dehydrated texture. An oven set as low as it can go, about 170°F (76°C) degrees, does the job nicely in about twenty-four hours or so, with intermittent stirring to guard against hot spots. It's a simple matter to grind completely dried roasted koji in a dry blender (or in very small batches in a coffee or spice grinder), and results in a powder that can be used as a sweetener or flour, albeit one with a very high sugar content. Because of the sugar content, it is a good idea to add a small amount of cornstarch or white rice flour to act as an anti-caking agent.

To use the powder as a flour, it's best to combine it with other flours, substituting no more than 25 percent of the total amount. The flavor is very strong and will come through even in small amounts. These rice flour snickerdoodles contain 12.5 percent roasted koji powder so as not to compromise the structural integrity of the cookies, and are rolled in the powder instead of the traditional cinnamon-sugar mixture. The powder on the outside caramelizes even further in the oven and adds an irresistible bittersweet quality.

**114 g unsalted butter, room temperature**
**76 g granulated sugar**
**60 g (2) egg yolks**
**5 g vanilla extract**
**174 g white rice flour**
**28 g roasted koji powder, plus more for coating**
**4 g baking powder**
**Pinch salt**

Preheat the oven to 350°F (176°C). Prepare a baking sheet lined with parchment or a nonstick silicone mat. (You may need to use more than one sheet, or save some dough to bake later if needed. It can be baked immediately after mixing but keeps well refrigerated and airtight for 2 to 3 days.)

With a hand mixer or in a stand mixer fitted with the paddle attachment, beat the butter and sugar until very uniformly fluffy. Scrape down the bowl to ensure there are no lumps, then beat in the yolks one at a time, scraping after each, followed by the vanilla extract. In a separate container, whisk together the remaining ingredients, then add to the batter and mix on medium-low to combine, finishing by hand if necessary. Drop dough by teaspoonfuls into a bowl of roasted koji powder and toss gently to coat, then arrange in a staggered pattern on the prepared sheet, leaving at least 1 to 2 inches (2.5–5 cm) between them. Bake for about 15 minutes, rotating the sheet as needed to ensure even color, until the outside edges of the cookies are firm and the koji powder has fully darkened and caramelized. These cookies are great as they are but also make lovely ice cream sandwiches, or can be ground into crumbs and used as the base for a miso cheesecake.

Yields 20 cookies

## Koji Breads

A number of years ago, the fantastic team at Tartine bakery in San Francisco started using koji in some of their bread recipes. Since then many people have worked to incorporate it into their loaves. You can now find everything from Asian-inspired milk breads to rye breads that are made using koji. One thing of note in bread making is that koji isn't used instead of a sourdough starter or yeast; it is used in harmony with them. Most notably, koji's role in the baking of bread—including those developed for Larder—is that of a flavoring agent and enhancer versus a microbial culture.

That's not to say you can't incorporate koji into your starter. Should you wish you can easily feed your sourdough starter with the sour amazake on page 115. Simply use the sour amazake instead of water and you're good to go. In any bread recipe, koji will work to not only enrich and fortify the flavors and tastes already present but also harmonize them beautifully.

# Amazake Rye Bread

For many years rye bread in America has been found mainly in the form of Jewish rye. While many people enjoy this bread, ourselves included, it barely has any rye in it! Most of the Jewish rye that you find nowadays contains no more than 5 percent rye flour. Deep, dark, richly delicious rye breads like those found in Ukraine, Scandinavia, and Germany are hard to come by. Thankfully there are more and more small-batch bakeries making rye breads that actually contain large amounts of rye. Rye is notoriously hard to work with, especially when your goal is to create a bread with a soft and pillowy crumb, as it is dense and heavy, and when used for bread can result in something with the texture of a hockey puck. While this is a style of bread that many people enjoy, there are just as many who want something light and fluffy.

We're not exactly sure why or how, but for some reason koji has the ability to lessen the density of a rye bread and allow you to create a light and fluffy bread even with a heavy ratio of rye. As time goes on we hope to get this bread into the hands of a scientist who can fully analyze it and tell us exactly what's going on. Until then we'll keep making this rye bread that Jeremy developed for Larder, which uses 50 percent rye flour yet eats like the light and fluffy Jewish rye breads that most of us are accustomed to. This bread uses the sour amazake from chapter 6 (see page 115) in its hydration, which serves to give the bread the pleasant tang that is found in many sourdough breads, without using a starter. At Larder this bread can more than stand up to the spicy mustard, sauerkraut, and bold pastrami in our signature sandwich. It also makes for an excellent cracker when the loaf is sliced thin and baked until crisp. A personal favorite for us is to slather it with butter and cream cheese and then top it with caviar.

**200 g water**
**200 g amazake**
**44 g molasses**
**3 g instant yeast**
**245 g rye flour**
**245 g high-gluten flour**
**8 g + 8 g wild carrot seed**
**8 g + 8 g caraway seed**
**8 g + 8 g mustard seed, equal parts yellow and brown**
**12 g + 8 g salt**
**8 g toasted yeast (for topping)**
**1 large egg**

In a mixing bowl, mix the water, amazake, and molasses. In a separate bowl, combine the yeast, flours, and the first portion of the spices and salt. Gradually stir the dry ingredients into the wet using a dough whisk or spoon until the flour is well incorporated. Cover with plastic and let rest for 15 minutes. After about 15 minutes, mix again for a minute or two. Again let rest for 15 minutes and mix one more time as before. Now cover the bowl with plastic and let sit at room temperature for roughly 12 to 14 hours.

After the long proof, stretch and fold the dough and shape it into boule or batard (round or torpedo) shapes for baking. Cover again with plastic and let rest for 15 minutes before putting in a proofing basket for the final rise. If you don't have a proofing basket, line a bowl with a well-floured kitchen towel and put the dough in there for the final rise. The final rise should last somewhere between 1 and 1.5 hours. Keep the dough covered with plastic to prevent it from drying out.

Preheat your oven to 350°F (176°C) half an hour before baking. Score the dough with a razor or sharp serrated knife. Beat the egg with water to make an egg wash. Mix the spices (second measurement) and yeast together and toast for a few minutes in a sauté pan over medium heat. Season the egg-washed dough with the toasted spices and yeast. Bake until the internal temp is about 200°F (93°C). Let cool completely before eating.

# Amazake Buttermilk Bread

Jeremy bakes this at Larder. It is made with the "Amazake-Cultured Cream" on page 215 that has been churned into butter and buttermilk. The flavors of the cultured dairy and the toasted koji are united in glorious harmony in this bread. It's great anytime you're looking for a classic American soft white bread. From peanut butter and jelly to tuna salad sandwiches, this bread will be your new go-to. The dough can also be formed into 4-ounce (113 g) balls and baked into a fantastically light and pillowy burger bun.

*For the bread:*
**119 g warm water**
**12 g sugar**
**16 g instant yeast**
**480 g bread flour**
**175 g spent grain koji**
**18 g kosher salt**
**43 g melted amazake butter, plus extra for pans**
**240 g amazake buttermilk**

*To garnish the loaf:*
**1 egg**
**22 g spent grain koji**
**9 g toasted yeast**
**6 g salt**

Mix the water and sugar. Bloom the yeast in the sugar water until it becomes frothy, about 10 minutes. Mix all ingredients except for the garnish in the mixer. Continue to mix and knead for a few minutes after the dough forms. The dough should now be slightly elastic and not sticky.

Proof the dough in a buttered bowl in a warm spot such as on top of your refrigerator until the volume doubles. Use your hands to gently compress the dough; portion, stretch, and fold it, then place in buttered loaf pans. Preheat the oven to 350°F (176°C). Proof until doubled again. Garnish the top of the loaf with an egg wash and sprinkle on spent grain koji, toasted yeast, and salt before baking. Bake until the internal temp is 195°F (90°C). Pull the bread out of the loaf pans and bake for an additional 3 to 5 minutes. Rest on a rack until it's cool enough to slice.

# ACKNOWLEDGMENTS

There are so many people that we would like to thank for making this book a reality. The community that has coalesced around using koji has been nothing but cosmic in its enthusiasm. Some of the people who were early champions of us and the exploration that we embarked on are Hallie & Eric Kogelschatz, Mary Redding, Savannah Jordan, Brett Oliver Sawyer, Dan Souza, Jonathon Sawyer, Jorge Hernandez, Lani Raider, Paul Wigsten, Nicco Muratore, Jeremy Kean, Peter Kim, and Geoff Lukas. Thank you for allowing us to investigate and blab. Thank you to everyone at our respective homes away from home for supporting us: the Larder family Allie La Valle, Kenny Scott, Angel Zimmerman, Alfred Sandoval III, Maximilian Schell, Valeria Flores-Villalon, and Katie Merchant for putting up with Jeremy complaining about deadlines and encouraging him to get back to work; and the Commonwealth Cambridge team for giving Rich whatever he needed regardless of how busy it was. We'd also like to acknowledge and thank Sandor Katz, Herve This, Alan Davidson, Elizabeth David, William Shurtleff, Akiko Aoyagi, Harold McGee, Ariel Johnson, and Dave Arnold for being pioneers in your respective works and inspiring us to continue on with ours. The impact that you all have had on us and countless others is immeasurable. Michael Harlan Turkell, Sandor Katz, Kirsten Shockey, Leda Meredith, and Matt Queitsch get mad props for helping guide us through the publishing world when we were shopping this book around. Thanks to Max Hull, Pete Larson, Andrew Wang, Claudia Mak, Matthew Claudel, and Billy Ritter for your amazing artistic talent. To Makenna Goodman and the entire team at Chelsea Green Publishing, thanks for seeing in this book what we do and for taking us into your family. Thank you to the following for giving us products to work with as we were developing and researching the methods and techniques in this book: Kate's Fish, Certified Angus Beef, Debragga Meats, Yellow House Cheese, Fallen Apple Farm, Ohio City Farm, Spice Acres Farm, Castle Valley Mills, The Buckle Farm, and Row 7 Seeds.

To the many contributors and others whom we mention in this book—Allie La Valle, Eric Edgin, Sam Jett, Kenny Scott, Diana Clark, Cynthia Graber, John Gibbons, Kevin Fink, Meredith Leigh, Kirsten Shockey, Stephen Lyman, Koichi Higuchi, Kim Wejendorp, Alex Talbot, Aki Kamozawa, Anna Markow, David Uyger, Jon Adler, Branden Byers, Sarah Conezio, Isaiah Billington, Kevin Farley, Akiko Katayama, Alex Hozven, Joshua Evans, Bentley Lim, Brian Kellerman, Mickey Kellerman, Mara King, Kristyn & Kevin Henslee, John Hutt, Irene Yoo, Harry Rosenblum, Coral Lee, James Wayman, Eugene Zelany, Brian Benchek, Alfred Francese, Sean Doherty, and Johnny Drain—we extend our deepest thanks and gratitude for making this book possible. Look to "Resources" at the back of this book to find out more about all of these amazing individuals and how you can follow their work.

---

First and foremost, I'd like to thank my best friend and soulmate, Allie, and my daughter, Emilia. You two are my bedrock, and I couldn't have done this without your support and understanding. Thank you to my parents, Dan and Joanne, and in-laws Franny, Brian, Bob, and Connie for championing my ideas and for helping out with childcare! Thank you to Kenny Scott, Angel Zimmerman, and Fred Sandoval for being such great sounding boards for so many of the ideas and their content in this book. Thank you to Noelle Celeste, Lisa Sands, and Kathy Carr for helping to foster my voice and prose by giving me a home on the pages of *Edible Cleveland*. Thanks to Jonathon Sawyer for letting me hit the ground running and championing my work, and to Judy Umansky, Graham Vesey, and Marika Shiori-Clark for helping to make Larder a reality. Thank you to all my family, especially my siblings Ethan, Rebecca, Julia, and Jon, and friends that have put up with eating all the weird things that I've concocted over the years. *Thank you, Cleveland*, and all the wonderful people who come to Larder to enjoy a meal! Very special thanks are due to my furry faithful companion and foraging buddy Baba Ganoush. You kept me company and made sure my feet stayed warm on many a late night after everyone else had gone to bed. I'd finally like to thank Richard Shih. I couldn't have had a better compatriot on this adventure.

—**Jeremy**

# ACKNOWLEDGMENTS

To Tanya, the love of my life, who has supported all of the craziness that comes along with pushing for change. I couldn't have gotten as far as I have today without her. To Maddy, my precocious daughter, who will forever ask all the right questions at the most inappropriate times. To my mother, who always makes sure I have everything I need no matter how old I get. To my father, who has encouraged me to always pursue my passions. To my brother, who has been cheering for me the whole time. And last, but not least, my brother from another mother, Jeremy, who has made me see that dreamers can make a difference.

—RICH

## APPENDIX A

# Fundamental Makes Quick Reference

During our workshops, we fill out a chart of ingredient ratios as we teach each koji application as a learning tool for participants. Guests have found this to be a handy quick-reference guide for whatever they make after the class. Table A.1 lists the major fundamental makes and base recipes. Once you have a basic understanding of each after having read the chapters and making your own, you'll appreciate how convenient this simple chart is to reference instead of having to flip to each recipe in each chapter.

**TABLE A.1.** Major Makes Quick Reference

| | Koji | Water | Cooked Starch | Protein | Salt to Total | Duration until Use |
|---|---|---|---|---|---|---|
| Amazake | 1 part | 2 parts | 1 part | N/A | N/A | 1–2 weeks |
| Shio koji | 1 part | 1 part | N/A | N/A | 5% | 1 week |
| Kojizuke base | 1 part | 1 part | 1 part | N/A | 5% | 1 week |
| Amino paste (ST) | 1 part | N/A | N/A | 1 part | 5% | 2 weeks–3 months |
| Amino paste (LT) | 1 part | N/A | N/A | 2 parts | 13% | 6–12 months |
| Amino sauce | 1 part | 2 parts | N/A | 1 part | 13% | 12 months |

*Notes:* ST = short-term; LT = long-term; "duration until use" assumes an ambient temperature environment; for a vinegar, wait on the amazake for at least two months; if the amino sauce koji is already a mix of starch and protein, exclude the protein and only add an equal part of water.

APPENDIX B

# A Deeper Dive into Making Aminos

For those interested in going deeper with aminos, this appendix is for you. If you're only just beginning your koji journey, you'll probably want to revisit this after around six months, when your pantry is being taken over by these tasty makes and you're looking to broaden your adventures. Here we will have friends of ours present to you two key factors of success: the science that drives the complex flavors and the technical details for optimizing your amino makes.

## Savory Science *by Bentley Lim*

Fermented foods, when compared with their starting products, are enjoyed for their richer and more complex tastes and odors. These sensory enhancements occur due to the metabolic activities of food-fermenting microorganisms, which determine food quality, generate flavor, and enhance palatability. Fermented soybean products, such as miso and soy sauce, are one of the most common fermented commodities worldwide and yield flavorful, savory, and aromatic compounds. Even though many other fermented foods yield similar compounds, much of the scientific research to identify the drivers of flavor development in fermented foods has focused around fermented soybean products due to their high demand and profitability in the Asian markets. In general the two main contributors to flavor compounds in fermented products are the microbes and the products these microbes ferment and break down. This section will focus on the contribution of microbes to flavor creation in soybean products, since the general principles in the development of aromatic and taste-active compounds will apply to many other fermented substrates, including charcuterie and cheeses.

Miso and soy sauce both provide a loaded source of flavorful aromatic compounds and molecules that impart an umami taste. Known as the fifth basic taste receptor, umami endows the sense of deliciousness and savory taste, and is associated with the palatability and pleasurability of food. *Umami* means "delicious" when translated from Japanese and has been described as "meaty," "bouillon-like," and "savory." Natural ingredients can produce an umami taste, such as from specific seaweeds (kombu), mushrooms, tomatoes, and dried fish (bonito). This led to the discovery of monosodium L-glutamate (also known as MSG) in the early 1900s, and now the taste of monosodium L-glutamate is used as an umami standard in sensory studies.[1] This typical umami taste is now recognized as one of the basic tastes—a statement confirmed by the finding of its taste receptor in mammals.

Monosodium L-glutamate is a combination of two molecules, the amino acid L-glutamate (in its acidic form) and sodium chloride (also known as salt). Acidic amino acids such as L-glutamic acid and L-aspartic acid have a sour taste on their own, but when associated with sodium ions from salt in a liquid solution, the molecules elicit an umami taste. Various amino acids (the building blocks of proteins) and their derivatives in their sodium salt form are capable of providing an umami taste. Additionally, other small molecules can do so as well. The best-known class of these are nucleotides, such as inosinate and guanylate, which also need to be in a sodium salt form to yield an umami taste.[2] While glutamate can be found in a variety of foods (meat, seafood, and vegetables), inosinate is found primarily in animal-based foods (fish and meat) and guanylate is most abundant in certain mushrooms. However, MSG is the standard for the umami flavor, and all other sodium forms of amino acids and their derivatives are usually less intense than that of MSG. The exceptions to this rule are the derivatives of *oxyglutamic acid* (found in some Japanese mushrooms), which can have an umami taste up to twenty-five times more intense than MSG.[3] Current research is still identifying and confirming other compounds (amino-acid-derived or not) that elicit an umami taste.

It is believed that taste and aromatic compounds in fermented products are formed in a general two-step sequence.[4] Initial digestion of the raw material (such as soybean) by enzymes (known as proteases) primarily produced by microorganisms or found in the raw material is followed by secondary enzyme degradation or chemical conversion of amino acids into derivatives. In the context of miso and soy sauce, the major contributor of the proteases is koji. Inoculation of rice or wheat with *Aspergillus* spores allows the fungal species to grow into a dense, metabolically active monoculture full of proteases and amylases that degrade proteins and starches, respectively. Adding koji to a mashed substrate allows these proteases and amylases to break down the raw material. Glutamic acid is the major component of soybean and wheat proteins, and the release of this amino acid is essential for the formation of the umami taste from fermented soybean products. Not all *Aspergillus* strains are created equal; several research groups have isolated strains that differ in their growth rates and their production of proteases, amylases, and other compounds that can affect the intensity of the umami taste in the final product as well as imparting differences in its sweetness (for example, increased production of amylases and saccharases), fruitiness, aromatic compounds, acidity (for example, citric acid production), and color.

During the second fermentation step of miso and soy sauce, a majority of the aromatic compounds are created.[5] When koji is combined with the raw material and salt, growth and metabolic activity of the *Aspergillus* fungus dramatically decreases. However, the abundant proteases and amylases in koji start working on the substrate, releasing nutrients and creating molecules to feed the next wave of microorganisms to ferment.[6] Much research on the taste of soy sauce has focused on these compounds and which microorganisms create them. The first significant wave of microorganisms consists of salt-tolerant lactobacilli bacteria, with similar species found in other fermented foods like cheese, yogurt, and sauerkraut. In the *moromi environment* (the solid mass of ingredients in a soy sauce or sake), the lactobacilli produce even more proteases to degrade the substrate into more umami-tasting compounds. Most important, lactobacilli species drop the pH of the product and potentially produce molecules to inhibit the growth of bacteria that would otherwise spoil the end product. As a result of the dramatic drop in pH, the lactobacilli bacteria are unable to grow. This event gives rise to the second significant wave of microorganisms, consisting of salt-tolerant yeast. These microbes carry out alcoholic fermentation on the sugars released from the wheat and rice in the koji and create the volatile aromatic and flavor components 4-hydroxy-2(or 5)-ethyl-5(or 2)-methyl-3(2H) furanone (sweet and caramel-like) and 4-ethylguaiacol (spicy, clove-like, woody, and smoky), creating a more full-bodied taste to the fermented end product.

The rate of formation and the quantity of these compounds in your final product will vary depending on the duration of the fermentation (three months to three years), salt content, temperature, and local community of microorganisms that are present when making your moromi. The salt-tolerant yeasts that contribute to these aromatic compounds have been identified, and several research groups have isolated yeast strains that grow faster and produce more of these volatile molecules, theoretically decreasing the time needed for aroma development in soy sauce.[7] However, the complex aromatic profile in each fermentation is highly dependent on the intricate dynamics among these microorganisms, and it is still unknown how the final composition of aromas will be affected.

The biology of umami perception is complex, and still, much of it is yet to be understood. Recently scientists have identified multiple umami receptors in mammals, and such activation of several receptors for umami compounds can elicit a more complex response for the consumer.[8] It is unclear whether many of the umami-tasting compounds identified so far interact with the same taste receptor for MSG. Interestingly, several umami-taste interactions have been identified. The best-documented is the interaction between inosinate or guanylate with MSG, which enhances the intensity and deliciousness of the umami flavor in a synergistic manner, to a point where inosinate supplementation can increase the sensitivity to MSG fifty-fold.[9] MSG is also known to interact with the perception of saltiness, enhancing the taste sensitivity to NaCl so that the use of salt in foods can be reduced. Lastly, researchers in 2015 found that the sensitivity to sweets was significantly decreased by MSG or other umami-active peptides but not by umami-active nucleotides, yet these nucleotides decreased the consumer's sensitivity to sourness and bitterness.[10] Further research will be important in uncovering more of these taste interactions to give consumers better control in making their products.

## Water, Time, and Temperature *by Sam Jett*

*The purpose of the following section by Sam Jett is to present basic ratios and guidelines for making amino sauces and pastes. The intent is to empower you with process understanding so you can make whatever you like. One recommendation is to be mindful that your final product is not only delicious but, more important, safe to consume. We can ferment just about anything, but that does not mean we should ferment everything.*

### THE IMPORTANCE OF WATER ACTIVITY

One factor that needs to be discussed in any deep dive on making aminos is water activity (referred to as $a_w$). This is the factor that will allow your ferment to stand the test of time. Since we are not doing ferments based on acidity, the control point on most liquid amino and miso is based on this factor. Active water is a percentage (partial vapor pressure divided by the standard-state partial vapor pressure of water) and is normally shown in a decimal point range (0.5–1.0, for example). Folks who are familiar with charcuterie will have a leg up on understanding this number and why it is so important as this is a critical control point for its production, for example, 30 percent water loss in hams.

Water activity is not a measure of how much water is in a product, but how much free water versus bound water there is in any given make. Water can be bound by sugar, starch, salt, and some hydrocolloids. Active water is the reason why peanut butter, jams and jellies, honey, and such are shelf-stable. Bacteria have trouble growing in products with an $a_w$ below 0.91, and fungus has trouble growing in products below 0.7 $a_w$. For the purposes of deciding how long we can ferment a miso or liquid amino, we need to factor in the control points first.

Measuring water activity is straightforward but requires the purchase of an expensive machine. For any commercial business that is going to sell this product, I highly recommend sending off products to process authorities and getting variances written. You are taking a risk if you don't (not to mention it is very illegal from the viewpoint of DHEC and the FDA). If you are a home cook, though, do not fret, as there are some easy ways to figure out water activity. For basic math, you can pretty much assume 10 percent of salt will bind 0.07 water.[11]

This means a product with 15+ percent salt (old-school fish sauce, for example) is going to be shelf-stable for a very long time and can be fermented and aged for several years. Something on the opposite side of the spectrum, such as a miso (around 4 to 6 percent salt), will have a lot of its water bound by the substrate and the koji.

Since the Province of Manitoba's agriculture website does a wonderful job mapping out the details, I've included their three water activity tables (B.1, B.2, and B.3).

**TABLE B.1.** Water Activity of Common Foods

| Foods | $a_w$ |
|---|---|
| Fresh meat and fish | 0.99 |
| Raw vegetables (ex: carrots, cauliflower, peppers) | 0.99 |
| Raw fruits (ex: apples, oranges, grapes) | 0.98 |
| Cooked meat, bread | 0.91–0.98 |
| Liverwurst | 0.96 |
| Caviar | 0.92 |
| Moist cakes (ex: carrot cake) | 0.90–0.95 |
| Sausages, syrups | 0.87–0.91 |
| Flours, rice, beans, peas | 0.80–0.87 |
| Salami | 0.82 |
| Soy sauce | 0.80 |
| Beef jerky | < 0.80 |
| Jams, marmalades, jellies | 0.75–0.80 |
| Peanut butter | 0.70 |
| Dried fruits | 0.60–0.65 |
| Dried spices, milk powder | 0.20–0.60 |
| Biscuits, chocolate | < 0.60 |

*Source:* "Water Content and Water Activity: Two Factors That Affect Food Safety," Province of Manitoba, 2019, www.gov.mb.ca/agriculture/food-safety/at-the-food-processor/print,water-content-water-activity.html.

Using the information in table B.1, we can make some more accurate decisions in putting together the ratios for the products we want to create. These guidelines do not mean your product won't spoil, but they will give you a leg up in preventing spoilage. Remember, with koji-based fermentation we are using the enzymes and proteases created by the death and autolysis of the *Aspergillus oryzae*. Our goal is to create circumstances that the culture on the koji dies so that their decomposition can get to work on the substrates we provide. I would aim for a lower water activity in any long ferments, such as aged amino pastes, and shorten the aging or fermentation on any ferments that are higher in $a_w$ (think wet amino pastes that you want to get a lot of tamari from). I do not recommend fermenting above 0.85 $a_w$; it is not safe

**TABLE B.2.** Typical Water Activity of Common Microorganisms

| Group of Microorganisms | Minimum $a_w$ for Growth |
|---|---|
| Most gram-negative bacteria | 0.97 |
| Staphylococcal toxin production (by *Staphylococcus aureus*) | 0.93 |
| Most gram-positive bacteria | 0.90 |
| Most yeasts | 0.88 |
| *Staphylococcus aureus* | 0.86 |
| Most molds | 0.80 |
| Halophilic bacteria (grow best at high salt concentrations) | 0.75 |
| Xerophillic molds (can grow on dry foods) and osmophillic yeasts (can grow in the presence of high concentrations of organic compounds, ex: sugars) | 0.62–0.60 |

*Source:* "Water Content and Water Activity: Two Factors That Affect Food Safety," Province of Manitoba, 2019, www.gov.mb.ca/agriculture/food-safety/at-the-food-processor/print,water-content-water-activity.html.

due to the risk that pathogens could be present and easily multiply, and especially with koji you will just end up with a boozy product.

Molds have minimum water activities for growth and toxin production. To produce mycotoxins, most molds require a higher water activity than the minimum requirement for growth. Table B.3 shows a few common mycotoxins and minimum water activities for mold growth and toxin production.

Make no mistake, the practice of making amino pastes and liquids can be considered a kind of controlled rot. We are using water activity, time, and temperature to ensure autolysis in a manner that does not allow for harmful microorganisms to grow. This helps us prevent the creation of toxins that not only make us sick but also will create off-flavors and -aromas. These factors are not the only variables that will give us a safe and tasty ferment, however. Cleanliness is critical as well as the product itself. We need to look at the substrate's composition to make safe decisions. Foods that are higher in fat tend to spoil (as all fats do when they oxidize) at certain temperatures. Some proteins spoil faster than others (fish versus

**Table B.3. Typical Water Activity of Common Molds**

| Mycotoxin | Mold | Minimum $a_w$ Requirement | |
|---|---|---|---|
| | | TOXIN PRODUCTION | GROWTH |
| Aflatoxins | *Aspergillus flavus*<br>*Aspergillus parasiticus* | 0.83–0.87 | 0.82 |
| Ochratoxin | *Apergillus ochraceus*<br>*Penicillium cyclopium* | 0.85<br>0.87–0.90 | 0.77<br>0.82–0.85 |
| Patulin | *Penicillium expansum*<br>*Penicillium patulum* | 0.99<br>0.95 | 0.81 |

*Source:* "Water Content and Water Activity: Two Factors That Affect Food Safety," Province of Manitoba, 2019, www.gov.mb.ca/agriculture/food-safety/at-the-food-processor/print,water-content-water-activity.html.

beans, for example). Being mindful of the substrate is paramount to creating a safe ferment. With higher fat or volatile proteins, a lower water activity is important.

## TEMPERATURE CONTROL FOR EFFICIENCY

The ability to control time and temperature is truly a cornerstone of modern food production. When we step back and look at the food industry, time and temperature are controlled at almost every step, from growing a product to serving it to a guest. So to me, it is no surprise that they are important factors in the control of a ferment. The basic idea is to think about temperature *in lieu of* time. If you can raise the temperature, then the protease and enzymatic reaction rate increases. If the rate of activity is increased, then the total ferment time can be lowered. Thus, temperature and time are inversely proportional (at least in theory). I certainly am not the first to realize this; I found some supporting arguments in a few science articles (dating back as far as the 1950s!) and I know the folks at Noma have been instrumental in pioneering this application of heat to ferments.

To me, this new way of framing time not only lets us speed up our koji-based ferments but also gives us the ability to influence their flavor. What

I mean by this statement is that we should use our temperature control to "cook" a ferment, just as we roast meat. When we get into the 140°F (60°C) range, we will slowly begin to caramelize the product. This helps it achieve a characteristic like aging that is associated with some ferments such as soy sauce. It is important to realize that if we are fermenting in the temperature danger zone (41–141°F / 5–60°C), we must monitor and tend to the product daily. Having the proper water activity and salt content will assist during the fermentation, but there are unsafe temperature ranges even if you have the proper controls in place to prevent spoilage. Please be mindful of this as we move forward through this discussion and please, if possible, test or send your product to a process authority to ensure its safety. Typically, when we lower salt and increase temperature we are building a product for speed, not endurance. When holding low-water-activity ferments after harvest, I recommend refrigerating them. This will help slow the fermentation process so there is not a loss of flavor or a growth of unwanted molds or yeasts. I do not recommend aging these types of ferments unless you are experienced in this process due to the lowered salt content.

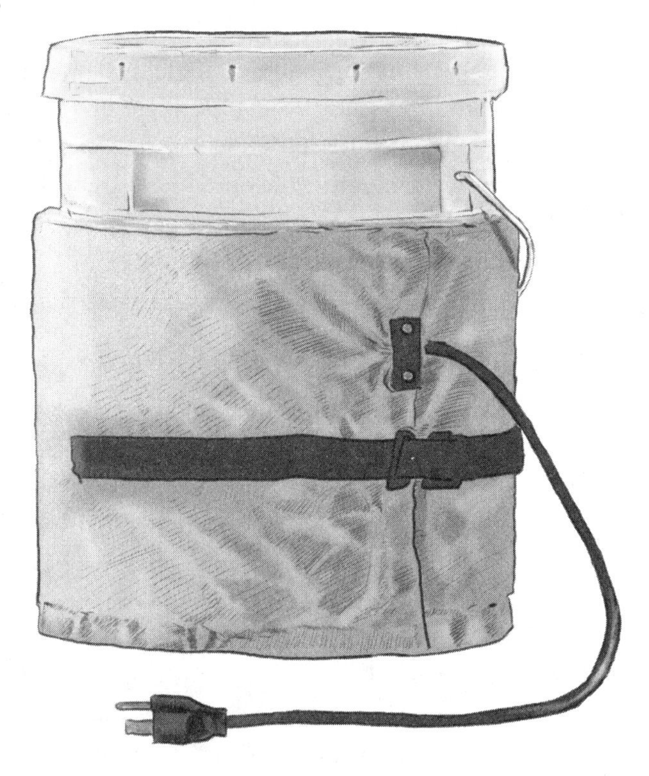

A simple heat acceleration option. *Illustration by Max Hull.*

I like to cook most miso from 91°F (33°C) (for products such as nuts that contain high fats that can spoil) to 109°F (43°C) (for low-fat products). I will run them at 4 percent salt for seven to fourteen days, or 6 percent salt for fourteen-plus days. They are normally ready to serve after this ferment time, no age needed. My rig for miso is a CVap, or immersion circulator. I pack products in stainless steel, non-reactive containers, place them in their chamber or bath, cover with plastic, and weigh down the top. This method was passed down to me from my predecessor at McCrady's, chef Josh Fratoni.

Most liquid amino products I will cook at 140°F (60°C) (or in some cases on a hot summer roof in Charleston, around 125°F/52°C). I tend to cook liquid amino sauces for thirty to sixty days, stirring often. If I am making them on the rooftop of my lab, then I will give them up to a year to age (this will get you to a traditional fish sauce style of caramelization), but you can pull them sooner. For example, country ham garum spends about two months on the roof in the summer, giving it a golden hue and a flavor like honey. I keep the last at an 8 percent minimum salt content and will scale it up based on the salt content of the substrate and length of the ferment.

I utilize industrial heating blankets used for heating solvents and glues. You can find them for less than $100 on Amazon; they cover crocks sized from 2 to 5 gallons (7.6–19 L). The best part of this move is that they hold at around 140°F (60°C) in a warm storage room without fail and can run for many months (the ones I have are still working after three years). It is recommended to stir ferments every day for the first week or so to ensure rapid temperature dispersion.

## TIME IS OF THE ESSENCE

I try to let my controlling factors of temperature and water activity decide how long I ferment or push the aging process. Just because your water activity is below 0.80 or your temperature is at 140°F (60°C) does not mean yeasts and thermophilic bacteria can't thrive. Anyone who ages miso has most likely come across a surface yeast. Most times it can just be skimmed off the top; the product underneath is fine. Finding these different nuances within your projects doesn't mean they are unsafe, just that they may need to be tended to.

Putting your ferments into temperature overdrive will help you work with a higher water activity. There are limits to this, of course, but it has helped me push the boundaries of the types of miso and liquid amino

**TABLE B.4.** Amino Make Guidelines

| Type of Ferment | Salt Content (%) | Time | Temperature | Suggested Substrates |
|---|---|---|---|---|
| Miso (high fat) | 4–6 | 2 weeks | 91°F/33°C | Low water activity; nut, dairy |
| Miso (low fat) | 4–6 | 2+ weeks | 109°F/43°C | Low water activity; legume, vegetable, mushroom |
| Liquid amino | 8–10 | 30 days | 140°F/60°C | Low water activity; country ham, vegetable, legume, mushroom |
| Liquid amino | 8–10 | 60 days | 140°F/60°C | Low to moderate water activity (.85 $a_w$ or below); meat, fish, dairy |
| Liquid amino | 12–15 | 6+ months | 68–77°F/ 20–25°C | Low water activity; dried products and grains |
| Liquid amino | 15–20 + | 1+ year | 68°F/20°C | Very low water activity; fish sauce |

products I will make. If I need or want to turn a product quickly or to create a more defined flavor, I will raise the water activity and the temperature. If I am aiming for depth and time is not an issue, then I will lower water activity and temperature and let it ride for a year or more. When making a ferment like fish sauce or a product with a volatile substrate, I will keep the water activity low for safety's sake. One plus of pushing the temperature and water activity of a product is that less koji is needed and a more recognizable flavor of the substrate is produced. In fact, it is rare that I go over 20 percent koji in my ferments to aid in the clarity of perceived substrate flavor in the final product. I know that is pretty low compared with most, but the lowered salt content and higher temperature make up the difference.

One last and very important reminder: Since this is a controlled rot, you get what you put into it. If you decide to clear out your freezer of old meat to make a beef garum, then your beef garum will taste like old freezer. Try to pick fresh and delicious products at their peak.

## MAKING AMINOS EFFICIENTLY

Fundamental understanding of the science and process really allows you to create delicious condiments with less failure. Understanding the conditions to maintain an environment so your amino doesn't go off is an important factor when you've invested products, effort and time. This feeds into the optimization of processing large batches for consistent end products. What we've shared here empowers you with technical knowledge and formulaic guidelines to give you a head start on success. Although you may not understand exactly what all of this means immediately, it will eventually become clear as you continue making more and more.

## APPENDIX C

# Visual Charts of Koji Make Relationships

These visual information maps are intended to help you understand the fundamentals of koji applications so it's easier to apply them across the board in all of your food making adventures. *Illustrations by Matthew Claudel.*

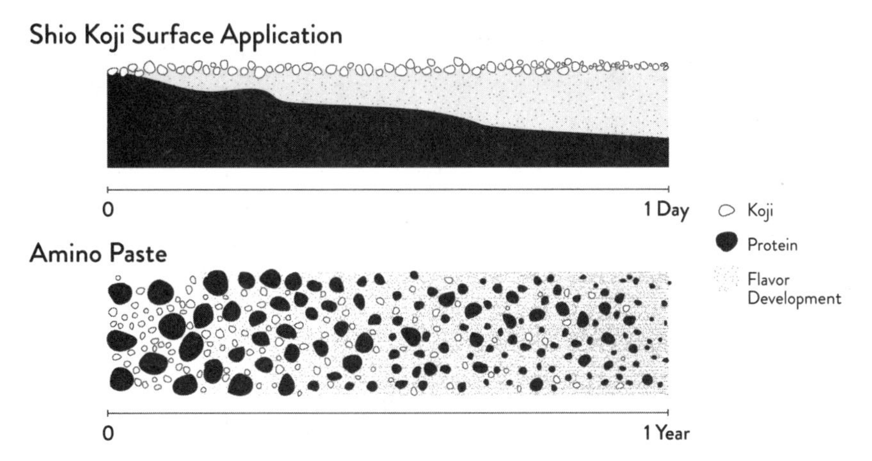


As we have discussed, there's a relationship between the short- and long-term koji applications. It's all about flavor development over time whatever that duration may be. For a shio koji surface application, you're quickly developing a kick of umami and sweetness to complement the base protein. For an amino paste, you're in it for the long run to develop complex flavors through an extended period of enzymatic breakdown and fermentation of the entire medium.

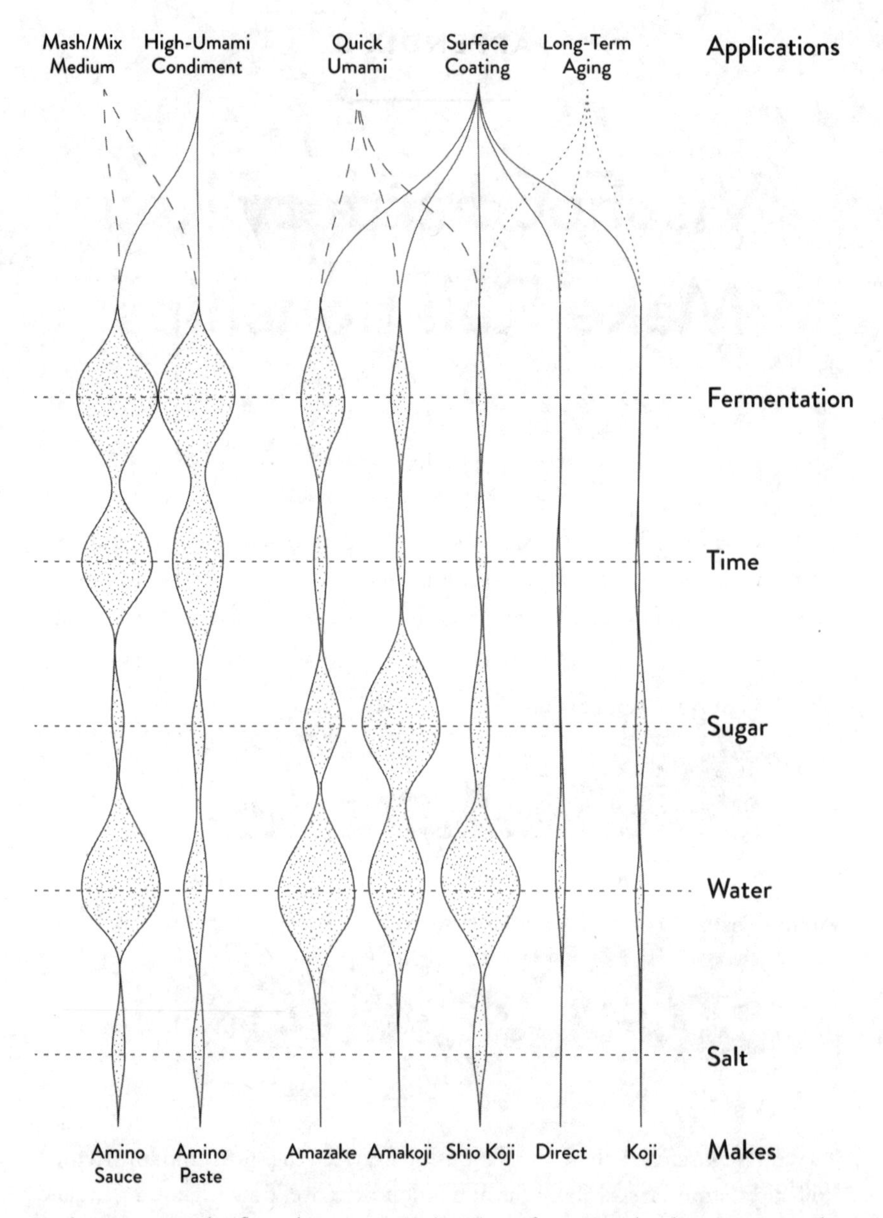


You've seen a similar flow chart on page 49. The information has been remapped to embed key components of the makes as waves with the width depicting ratios relative to other makes. We hope this helps you understand the relationships between them and why each may be better suited for one application over another.

APPENDIX D

# Food Safety

Established food safety rules lead to many questions about how to safely grow a mold intended for human consumption. Koji is a foreign substance to many people, especially those who work to manage and implement the enforcement of food safety laws. This being the case, the more you know about koji's precedent use, behavior, and growth requirements, the better equipped you'll be to deal with your local regulators in the event that they have questions. By providing an understanding of proven practices and basic microbiology in this appendix, it is our hope you will be comfortable with identifying the differences between good and bad practices with koji. We will address food safety for home and professional cooks, as well as address the additional steps required for food service professionals.

Although the subject of food safety can be dry and boring to discuss, it is quite possibly the most important thing to have a solid working knowledge of when producing koji and fermented foods, or any food for that matter. It is your responsibility to ensure that any food you produce or cook for others is safe. Unfortunately, many people casually cook dinner or experiment with food at home without understanding the risks. Professional culinarians are considerably more mindful of food safety due to intense and often burdensome regulations, but as we stated before there really isn't anything more important when it comes to preparing and serving food. Pathogens, bacteria, viruses, and parasites that can cause illness and disease are a real concern; anything that can be done to eliminate and mitigate their effects should be done.

It should be noted that while we have done our due diligence to vet this information and present it to you, different rules, regulations,

## A NOTE ON FOOD SAFETY

### Brian Kellerman (HACCP, PCQI, SQF, BRC, and ISO 9001 certified)

*Brian and Mickey Kellerman of Kellerman Consulting in Columbus, Ohio, are food safety specialists who work with food-based businesses to develop food safety plans that are in accordance with their specific local regulations. These are called HACCP plans (see appendix E for an example) and address all manner of concerns related to what could potentially go wrong during the production of a food and how to manage and minimize the inherent and apparent risks involved. Brian spells out exactly why food safety should be taken seriously.*

Food safety is beautiful. The kitchen is a place for experimentation, artistry, alchemy, and boundary pushing. It is a place where all of us are able to interact with ourselves and satisfy our needs and desires with the basic and complex flavors of life. As we control the natural and artificial processes that make pungent, delicious, and extraordinary culinary creations, the tools of measurement, precision, and control allow us to challenge ourselves, to create and enjoy our sustenance and share it with those special to us.

But food can also be dangerous. Whether you are an artist, a scientist, or just plain hungry . . . care, cleanliness, and safe practices are a must. Food poisoning is not always confined to an unpleasant night next to your commode. Crohn's disease and other chronic illness sometimes follow bouts of listeriosis or salmonellosis. Hemolytic uremic syndrome can destroy our kidneys and at its worst is fatal. As a food safety expert, my hope for anyone that truly loves food is to try everything and enjoy every meal. Experimentation and fearlessness are requirements for great chefs, but taking unnecessary risks out of carelessness or lax attitudes towards food safety is not worth it.

Put simply: Know your food safety. Whether the cooking techniques you use such as grilling, baking, or frying makes your food safe, or cold storage temperatures assures safety, the only way to know for sure is by *using a thermometer every time*. For foods that are shelf-stable, there are measured characteristics that determine safety, and it is your responsibility to know what those are. Fermentation and dry curing are excellent food preservation and food safety techniques that have been proven to work for thousands of years, but measuring the pH with calibrated pH strips and measuring water activity with a water activity meter actually tells us for sure that these foods are safe.

The ingredients, equipment, utensils, and containers we use to prepare our dishes should be organized, easy to find, easy to see, and easy to clean. We need to clearly communicate the rules and expectations of the kitchen with each person we share the kitchen with by properly labeling, and we need to record notes and observations in the kitchen so that we stay on top of changes and improvements. Cooking and recipe development are a requirement for the evolution of food and the discovery of wonderful and novel foods. Every cook is capable of pushing the boundaries of culinary excellence forward, but no chef should do so without care and safety. Cooking is beautiful, food is beautiful, and safety is beautiful.

and data exist depending on where you live. The rules and regulations we present here reflect where we live (Cleveland and Boston, United States); our food safety laws and best practices differ from those in New York City, which in turn are different from those just a couple of hours away in Philadelphia let alone Mexico City, Rome, or Tokyo. It is your responsibility to ensure that what you are doing with koji, or any other food, is in line with your local regulations and laws. This holds true for everyone, professional and amateur alike. Now let's walk through the food safety basics that will ensure the foods you make using koji will be safe to consume.

## Cleanliness

"Be clean and work clean" is a motto etched into the mind of every culinary school student across the world, and it all starts with personal hygiene. Things that many people do regardless of whether or not they work in a professional kitchen are often some of the most important aspects of food safety. While some of the items on the following lists may seem like common sense, they aren't always at the forefront of everyone's mind when they step into a kitchen.

When we say, "Be clean," we are referring to the following:

- Keep your body, especially your hands, clean.
- Don't work with food if you are sick. This includes everything communicable from a cold to the stomach flu.
- Don't work with food if you have the symptoms of being sick but don't *feel* sick. This includes everything from an unexpected runny nose to diarrhea. Just because your whole body doesn't feel sick doesn't mean that you aren't carrying a pathogen that can sicken others. Everyone's body reacts differently to different pathogens.
- Keep your head and body hair short. If it's long then keep it tied back away from the front of your body and covered. This includes facial hair!
- Keep your fingernails trimmed and clean. You'd be surprised and disgusted if you looked under your fingernails with a magnifying glass no matter how clean you are.
- Wash your hands anytime you touch your face or any other part of your body.
- Wear clean clothes and keep them clean.
- Keep open wounds clean and covered. This applies to everything from the seemingly small and unnoticeable paper cut to an infected hangnail or a suture.
- Keep your bodily fluids inside your body. We all sneeze, cough, and bleed. If you do any of these things when working with food, however, then discard any food that came in contact with said fluid and thoroughly clean any work surfaces and areas of your body that came in contact. This also applies to solid pieces, including strands of hair that fall off your body or pieces of you that may be accidentally removed. Fingertips do get cut off from time to time.

When we say, "Work clean," we are referring to the following:

- Don't cross contaminate your work surface. This means that when you are working with raw or cooked meats on a cutting board, you should clean it appropriately before cutting herbs or vegetables on the same cutting board. It also applies to tools and utensils that you may be using. Don't stick a spoon in your amino paste and then into your amino sauce without properly cleaning it.

- Keep your work surface clean of debris and clutter.
- Keep your utensils clean.
- Stay organized and keep your ingredients properly stored.
- Your cutting board isn't your only work surface. Keeping the table around and underneath your cutting board clean is just as important.

## How to Wash Your Hands

Proper handwashing technique is often the first line of defense when it comes to keeping food safe. But simply rinsing your hands under running water is not effective enough to truly clean them. The amount of time that you wash your hands and the temperature of the water are also important factors. Professional culinarians wash their hands countless times during the course of a work shift, and you should adopt this practice as if you were working in a professional kitchen. The basic tips for handwashing are:

- The water you use should be as hot as you can handle.
- Use soap. According to the US Food and Drug Administration, there isn't enough science to show that over-the-counter antibacterial soaps are better at preventing illness than washing with plain soap and water (FDA Consumer Update, updated November 6, 2017).
- Wash and scrub your hands for a minimum of twenty seconds. This includes under your fingernails.
- Rinse off the soap.
- Turn off the water with a paper towel. Discard this towel.
- Dry your hands with a new paper towel or hands-free device.

## The Danger Zone

Now that we've addressed personal hygiene, we want to discuss other factors that can cause a foodborne illness to take root. The first and most important thing to note is what is referred to as time and temperature abuse, which happens when foods are left in what's commonly referred to as the danger zone for too long (any temperature between 41 and 140°F / 5–60°C). Within this range it is possible for most pathogens to multiply to a density that will allow them to infect and sicken someone. Foods such as potato salad or grilled meats should not be left out at this range for more than four hours. Exceptions are foods high in salt or acid, those low in available water, and those undergoing long-term cooking that will effectively pasteurize them after a predetermined amount of time. Fermented foods, koji, properly canned foods, and charcuterie are examples of the previously mentioned foods that can be safely kept in the danger zone. Simply put, keep cold foods cold and hot foods hot.

A properly calibrated thermometer is your best friend when questions arise as to whether food has been subject to time and temperature abuse. Follow the manufacturer's guidelines for ensuring that your thermometer is properly calibrated. True temperatures are obtained from whole, intact foods by inserting the probe into the center of the food.

APPENDIX E

# Dry-Cured Animal Foods HACCP Plan

This is an early version of the HACCP (Hazard Analysis Critical Control Point) plan that Jeremy developed to use at Larder. Nowadays HACCP plans are being requested by more and more regulatory bodies, and there is a lot of confusion as to what they are and how to use them. It should be noted that these are only used by people who are producing food for public sale and consumption, and are of no concern for any home cook. HACCP plans work to identify potential hazards during the course of the production of a food that can become contaminated with pathogens or materials such as hazardous chemicals or objects. They were originally devised to hold large industrial food producers accountable for life-threatening mistakes that can occur during the production of the foods that they make. Currently many jurisdictions require these plans not only for large industrial food producers but also for small restaurants and producers who make foods that have been identified as potentially hazardous. It isn't industrialization that is solely liable for food poisoning or hazardous contamination, for this can happen in any setting where food is produced, be it a factory or your favorite neighborhood restaurant. These foods include pickles, smoked foods, sushi rice, vacuum-packed foods cooked sous-vide, and charcuterie amongst others.

You should note that this is an *example* of an HACCP plan for you to see what one looks like and to get a greater understanding of the work you will need to do to not only develop your own plan but to also actively use and follow the plan you create. It is our experience that many professional cooks have hardly seen an HACCP plan yet alone actively developed and worked with one. This HACCP plan solely serves as a guide and is not designed to be used and implemented.

## Not-Heat-Treated, Dry-Cured, Not-Shelf-Stable, Whole-Muscle Animal Foods

This plan complies with the statutes found in the 2016 Ohio Uniform Food Safety Code including §3717-1-03(J)(2) for the variance requirement for the curing of foods in addition to §3717-1-03.2, §3717-1-04.5, and §3717-1-04.6 regarding prevention of cross contamination, cleaning, and sanitization procedures. An HACCP plan is required as part of the written approval process for the dry curing of animal foods in place of heat treatment to prevent any potential hazards. This plan is intended for the in-store production and storage of dry-cured animal foods for sale to customers. Actions described in this document are to be performed by trained staff only, and all items cured per this plan must be kept segregated from raw ingredients at all times before, during, and after the curing process. All records related to activities found in this document are to be stored for a minimum of two years and easily accessible upon request from a regulatory inspector.

### GENERAL HACCP INFORMATION

| | |
|---|---|
| **The HACCP Team:** | Jeremy Umansky, owner |
| **Product Description:** | Dry-cured, whole-muscle animal foods (beef, chuck eye) |
| **Ingredients:** | Beef, chuck eye, salt, Cure #1*, Cure #2*, *Aspergillus oryzae*, rice flour |
| **Hazards:** | Biological: Naturally occurring pathogens such as *Salmonella* spp., *Escherichia coli* 0157:H7, *Listeria monocytogenes*, *Staphylococcus aureus*, *Trichinella spiralis* |
| **Materials:** | Cryovac bac, packaging and labeling materials |
| **Equipment:** | Calibrated scale, VacMaster chamber sealer, refrigerator/freezer |
| **Labeling Instructions:** | Keep refrigerated or frozen |
| **How Is It to Be Used?** | Consume as purchased |
| **Where Is It to Be Sold?** | Retail |

* Cure #1 and Cure #2 are used separately and call for differing dry times.

## Sanitation Standard Operating Procedures

Only kitchen employees trained on this plan may handle food and equipment used as part of the dry-cure process. Before handling dry-cure ingredients or equipment, the trained kitchen employee must thoroughly wash hands and make sure the station has sanitized utensils and sterile gloves available. Bare-hands touching any ingredients (raw or ready-to-eat) as a part of this plan is not allowed during any step of preparation. Sterile gloves should be worn during the processing of food under this HACCP plan. The preparer must change gloves as frequently as is required to prevent contamination of foods during preparation and handling of foods. Hands must be washed prior to donning new gloves, and sanitized utensils must be used to transfer foods from preparation, curing, reduced oxygen packaging, inoculation, incubation, and drying areas.

Preoperative checks of all materials, ingredients, utensils, and food contact surfaces used in conjunction with this HACCP plan must be performed by trained staff before the beginning of production each day when dry-cure processing of animal foods occur for selling or shipping to customers. Observations may be recorded on the Larder *Sanitation Preoperational Record Form*. The observer must verify that all mechanical parts of the reduced oxygen packaging machine and any other equipment used within this process were disassembled (if possible) before cleaning and cleaned according to recommendation from the manufacturer. The observer must also verify that all utensils and food contact surfaces are unsoiled and were sanitized using approved sanitizers and concentrations. All equipment, food-contact surfaces, and utensils associated with the dry cure will be clean to sight and touch and should be cleaned and sanitized:

1. Between handling of different types of meat
2. Between change in working with meat and non-meat products
3. Between changes in working with raw foods and ready-to-eat foods
4. Before using or storing a food temperature measuring device
5. At any time in which contamination may have occurred
6. After final use each working day

Reduced oxygen packaging occurs in a designated preparation area that is not used for any other food processing.

## METHODS

In compliance with §3717-1-04.5, and §3717-1-04.6 of the 2016 Ohio Uniform Food Safety Code 2009, washing and sanitization schedules are based on the type and purpose of equipment or utensil, type of food involved, amount of food residue accumulated, and temperature at which the food is maintained during the operation and potential for growth or multiplication of pathogenic or toxigenic microorganisms. Washing procedures are selected by the type and purpose of the equipment or utensil, and on the type of soil to be removed. Equipment, utensils, and food-contact surfaces can be effectively washed to remove or completely loosen soils using manual or mechanical means and multiple methods such as application of approved type and concentration of detergents, emulsifiers, and other wetting agents; hot water, brushes, scouring pads, high-pressure sprays, and ultrasonic devices. Larder uses the following cleaners and sanitizers within their facility [Note: Types and brands have been omitted from this book]:

| Type | Brand | As Is or Diluted? |
|---|---|---|
| | | |
| | | |

At Larder, all sinks are provided with hot and cold water, self-draining drainboards, and easily removable dish tables. Sinks and drying areas are constructed so that soiled and cleaned utensils and equipment are kept entirely separate and protected against contamination from soiled utensils or dishwashing operations. Equipment, utensils, and food-contact surfaces with visible food debris are scraped or, if applicable, preflushed, presoaked, or scrubbed when necessary prior to the manual washing or warewashing for equipment and utensils.

For the warewashing of equipment and utensils, soiled items must be loaded onto racks, trays, or baskets in such a way that exposes the items to the unobstructed spray from all cycles and allows items to drain. For equipment in which washing in sink compartments or in a warewashing machine is impractical, such as fixed equipment (that is, reduced oxygen packaging machine) or large utensils, washing may be completed using an alternative manual warewashing method. Equipment must be disassembled as necessary to allow detergent to reach all parts, scraped or rough-cleaned to remove particle accumulation, and effectively cleaned to remove or completely loosen soil. The equipment manufacturer's instructions for cleaning should be referred to during this process. All methods of cleaning, including manual, warewashing, and alternative warewashing, must include a rinsing procedure that allows for the removal of abrasives and cleaning chemicals. Cleaning is followed by a sanitization step, which may include a hot-water or chemical sanitizing method as written in §3717-1-04.6(C).

## Standard Operating Procedure

In order to assure safe food to customers using curing as a food safety component, it is necessary to receive animal foods from approved sources (USDA grade), and all processing steps must be handled so as to render the food incapable of supporting the growth of pathogenic microorganisms prior to selling the product to customers. Due to the risks associated with the dry-curing process, it is important that bare-hand touching is prohibited in the curing area as well as throughout all steps of this plan, and that all procedures are strictly followed so as to assure proper penetration and distribution of curing into animal foods. The following procedures must be followed:

1. Raw animal foods must be received and stored at refrigerated temperature (≤ 41°F/5°C).
2. Raw animal foods should be brought out of refrigeration in small batches and on a first-in first-out (FIFO) basis in order to prepare items in such a manner that prevents them from reaching temperatures above 44°F (7°C) prior to packaging step.
3. In between preparation and packaging of one animal food to another, the designated reduced oxygen packaging area must be fully cleaned and sanitized as described in the Sanitation Standard Operating Procedure section.

4. Raw animal foods are processed, cut, trimmed, or de-boned. Raw animal foods are weighed or metered, which allows for calculation of the appropriate amount of curing ingredients to prepare.
5. Ingredient Dry Storage (Critical Control Point [CCP] 1): During removal of curing ingredients from storage and preparation of cure, ingredients must be visually inspected for any wetness as this could impact the effectiveness of the cure. Record on the Larder *Dry-Cured Animal Foods HACCP Monitoring Log.*
6. If curing ingredients are observed as wet, they must be discarded and recorded on the Larder *Deviation Corrective Action Record.*
7. Application of Cure According to Recipe (CCP 2): Curing ingredients are prepared, scaled, and rubbed onto all parts of the animal food according to recipe: 2.5% salt and 0.25% Cure #1 to 100% non-poultry animal food; or 3% salt, 1.5% sugar, and 0.25% Cure #1 to 100% poultry. Record on the Larder *Dry-Cured Animal Foods HACCP Monitoring Log.*
8. Animal food with applied cure is then vacuum-sealed as pressure allows for full penetration of cure. Vacuum-packaged animal foods are labeled with packaging date, instruction to maintain at ≤ 41°F (5°C), and to discard within 30 calendar days if not used or consumed. Record on the Larder *Dry-Cured Animal Foods HACCP Monitoring Log.*
9. Packaged animal foods are immediately transferred to refrigeration (≤ 41°F/5°C) for cold holding.
10. Cold Holding (CCP 3): Per Larder *Dry-Cure Recipe*, curing animal food remains in vacuum packaging for an amount of time that equals 1 day per 0.5 inch (13 mm) of muscle measured from the center of the thickest part of the muscle, in addition to an extra 24 hours. After this amount of time has elapsed, the animal food is removed from the packaging, transferred to a clean open container, and allowed to sit uncovered in refrigeration for an additional 36 hours. Record on the Larder *Dry-Cured Animal Foods HACCP Monitoring Log.*

11. Prior to mold inoculation, the coating mixture is scaled and combined according to recipe.
12. Cured animal food is removed from cold hold after allotted amount of time and is coated with a prepared mixture of *Aspergillus oryzae* (5 ml per kg of meat / 0.08 ounce per pound) and rice flour (120 g per kg of meat / 1.92 ounces per pound).
13. Incubation (CCP 4): The inoculated meat is incubated at 85°F (29°C) and 90% RH for 36 to 48 hours until mold has fully colonized the surface of the muscle. Record on the Larder *Dry-Cured Animal Foods HACCP Monitoring Log.*
14. After the incubation period, the inoculated animal food is weighed and hung for drying.
15. Drying (CCP 5): Drying process occurs at 55°F (13°C) and 85 to 95% RH until it reaches a weight loss totaling 35%. Record on the Larder *Dry-Cured Animal Foods HACCP Monitoring Log.*
16. If Cure #1 was used, drying process typically lasts ≥ 14 days, and if Cure #2 was used, drying process typically lasts ≥ 45 days.
17. Dry-cured animal food is weighed after the drying process, and if it has reached the goal weight loss of 35%, it is packaged and labeled for in-store sale to customers.
18. The finished product is packaged, labeled, and transferred to cold storage until in-store sale to customers.

## Ongoing Verification

The Larder verification process utilizes the HACCP Verification form for all ongoing verification that the facility and cooking equipment are functioning properly to achieve the necessary conditions to assure safe product for consumers. Calibration is observed and recorded on the *Weekly Calibration Record*, and will be performed for each of the following CCP equipment:

- Dry bulb thermometer
- Relative humidity gauge

Calibration of thermometers will be performed in accordance with the University of Nebraska Thermometer Calibration detailed below. Relative humidity will be calibrated through use of a kit that complies with ASTM standard E104-85.

A documented preventive maintenance is performed regularly to assure that the equipment is in continual condition to meet the required temperatures to meet the requirements in this HACCP plan.

Each batch of product will be held for a complete review of production records. Upon confirmation of the conformity to HACCP plan, product can be released from the facility. Any non-conforming finding or incomplete records will render the production batch out of conformity with the HACCP plan and unfit for human consumption. Non-conforming product will be quarantined by the HACCP team leader and disposed of in accordance with state regulation.

## Thermometer Calibration

HACCP-based food safety programs require accurate record keeping to be successful. Temperature is often the parameter of interest when monitoring a CCP. To assure that a temperature-dependent process is under control, a calibrated thermometer must be used to record temperatures. The majority of thermometers can be calibrated following a few basic procedures.

To be considered accurate, a thermometer must be calibrated to measure within +/– 2°F (1°C) of the actual temperature. Actual temperature can be determined in a variety of ways including measurement with an NIST (National Institute of Standards and Technology)-certified reference thermometer or simply through using an ice-water solution or boiling water. Another option is the use of sophisticated, and often high-cost, calibration equipment that is increasingly becoming available commercially.

The simplest and cheapest way to calibrate a thermometer is through the use of either ice water or boiling water. Distilled water should always be used, as dissolved solutes in tap water can significantly affect both freezing and boiling points. Another important consideration is the altitude (table E.1 on page 296) at which calibration is being performed. At sea level, pure water boils at 212°F (100°C), but at 10,000 feet (3,048 m) above sea level, it boils at only 194°F (90°C). Barometric pressure also has an effect on boiling point, but the effect is much less than that of altitude. You can visit WorldAtlas.com to determine the altitude of your city.

Thermometers intended for measuring higher-temperature items, such as cooked products, should be calibrated in boiling water while those used for taking lower temperatures should be calibrated in ice water. When calibrating in ice water, both the water and ice should be composed of distilled water. In either case care should be taken to prevent the thermometer from contacting the container being used, as this could result in erroneous temperature readings.

### CALIBRATION IN ICE WATER

1. Add crushed ice and distilled water to a clean container to form a watery slush.
2. Place a thermometer probe into the slush for at least 1 minute, taking care to not let the probe contact the container.
3. If the thermometer does not read between 30° and 34°F (−1 to 1°C), adjust to 32°F (0°C). Non-adjustable thermometers should be removed from use until they have been professionally serviced.

### CALIBRATION IN BOILING WATER

1. Bring a clean container of distilled water to a rolling boil.
2. Place a thermometer probe into the boiling water for at least 1 minute, taking care not to let the probe contact the container.
3. If the thermometer does not read between 210° and 214°F (99–101°C), adjust to 212°F (100°C). Non-adjustable thermometers should be removed from use until they have been professionally serviced.

Thermometers that are found to be inaccurate (that is, do not measure within +/− 2°F/1°C of the actual temperature) should either be manually adjusted or serviced by a professional. Thermometers that have a history of deviating from actual temperature measurements should be discarded and replaced. To assure accuracy, NIST-certified thermometers must be recertified annually.

Thermometers that cannot be easily calibrated through direct immersion in boiling or ice water can be calibrated by comparing readings with another calibrated thermometer. Thermometers that may be calibrated in this way include smokehouse probes and room temperature thermometers. When doing this it is important that the thermometer used for the comparison has been calibrated recently. All thermometers should be calibrated regularly, with those used for monitoring CCPs being calibrated either daily or weekly, depending on the volume of your operations. Any thermometer that has been subjected to abuse, such as being dropped on the floor, should be immediately recalibrated to assure accuracy. Hard-to-calibrate thermometers could be compared directly with NIST reference thermometers, but this may be undesirable as many of these reference thermometers are glass and mercury and could present chemical and physical hazards in food production areas.

**TABLE E.1.** Relationship of Altitude to Boiling Point of Pure Water

| Feet Above Sea Level | Boiling Point | Feet Above Sea Level | Boiling Point | Feet Above Sea Level | Boiling Point |
|---|---|---|---|---|---|
| 0 | 212°F | 3,000 | 206°F | 7,000 | 199°F |
| 500 | 211°F | 3,500 | 205°F | 8,000 | 197°F |
| 1,000 | 210°F | 4,000 | 204°F | 10,000 | 194°F |
| 1,500 | 209°F | 4,500 | 203°F | 12,000 | 190°F |
| 2,000 | 208°F | 5,000 | 203°F | 14,000 | 187°F |
| 2,500 | 207°F | 6,000 | 201°F | | |

*Source:* © 2005 University of Nebraska-Lincoln, Institute of Agriculture and Natural Resources, Lincoln, NE 68588

## Process Flow Reassessment

Reevaluation of the HACCP plan, including use of the initial validation forms for a ninety-day period (minimum thirteen production runs), must be used for any changes to the HACCP plan or introduction of new products to this HACCP plan. The continual use of verification documents is required for the life of the HACCP plan, and the HACCP team will review this HACCP plan and the production system at least annually, taking into consideration identified risks to the system, non-conformities in production, as well as verification data taken over the previous time frame under review. The HACCP team must document all reassessments of this HACCP plan with said documentation indicating the reason for the reassessment, and an explanation for the findings of the reassessment. If the results of any reassessment of the HACCP plan results in unsatisfactory outcomes, the HACCP plan must undergo revision in both the Design (Element 1) and Execution (Element 2) of the HACCP plan, and a new validation of the process, followed by continued verification of the effectiveness of the changes made to the system.

## Process Categories and Ingredients

**Process Category:** Not-Heat-Treated, Dry-Cured, Whole-Muscle Animal Foods

**Plant Name and Location:** Larder

| MEAT BY-PRODUCTS | NON-MEAT INGREDIENTS | BINDER EXTENDERS |
|---|---|---|
| Whole-muscle cuts of beef, chuck eye | Salt, sugar, rice flour, *Aspergillus oryzae* | See labeling for further detail |
| **SPICES, FLAVORINGS** | **RESTRICTED INGREDIENTS** | **PRESERVATIVES, ACIDIFIERS** |
| [Note: List spices/ flavorings per your recipe] | Cure 1: contains sodium nitrite<br>Cure 2: contains sodium nitrate | See labeling for further detail |
| **ALLERGENS** | **WATER** | **PACKAGING** |
| None | Must be potable | Approved packaging materials |

**TABLE E.2.** Flow Diagram

| Ingredient/Processing Step | Food Hazard | Reasonably Likely to Occur? |
|---|---|---|
| **1. RECEIVING AMBIENT NON-MEAT INGREDIENTS** | BIOLOGICAL—Vegetative pathogens such as *E. coli* O157:H7, *Salmonella* ssp., *L. monocytogenes*, *S. aureas*, *T. spiralis* | No |
| | CHEMICAL—Contamination from shipping vehicle, sodium nitrite, sodium nitrate | No |
| | PHYSICAL—Contamination from shipping vehicle | No |
| **2. RECEIVING RAW ANIMAL FOODS AT ≤ 41°F (5°C)** | BIOLOGICAL—Vegetative pathogens such as *E. coli* O157:H7, *Salmonella* ssp., *L. monocytogenes*, *S. aureas*, *T. spiralis* | No |
| | CHEMICAL—Contamination from shipping vehicle | No |
| | PHYSICAL—Contamination from shipping vehicle | No |
| **3. RECEIVING PACKAGING MATERIALS** | BIOLOGICAL—None | No |
| | CHEMICAL—Contamination from shipping vehicle | No |
| | PHYSICAL—Contamination from shipping vehicle | No |
| **4. INGREDIENT DRY STORAGE (CCP 1)** | BIOLOGICAL—Vegetative pathogens such as *E. coli* O157:H7, *Salmonella* ssp., *L. monocytogenes*, *S. aureas*, *T. spiralis* | Yes |
| | CHEMICAL—Allergens present in facility, sodium nitrite/nitrate | No |
| | PHYSICAL—None | |
| **5. COLD STORAGE OF RAW ANIMAL FOODS AT ≤ 41°F (5°C)** | BIOLOGICAL—Vegetative pathogens such as *E. coli* O157:H7, *Salmonella* ssp., *L. monocytogenes*, *S. aureas*, *T. spiralis* | No |
| | CHEMICAL—None | |
| | PHYSICAL—None | |
| **6. STORAGE OF PACKAGING MATERIALS** | BIOLOGICAL—None | |
| | CHEMICAL—None | |
| | PHYSICAL—None | |

| Basis | If yes to column 3, what measures could be applied to prevent, eliminate, or reduce the hazard to an acceptable level? | Critical Control Point |
|---|---|---|
| Supplier approval policy, truck inspection, letters of guarantee | | No |
| Supplier approval policy, truck inspection, restricted ingredient log | | No |
| Supplier approval policy, truck inspection | | No |
| Supplier approval policy, truck inspection, letters of guarantee from suppliers addressing BSE and pathogen control | | No |
| Supplier approval policy, truck inspection | | No |
| Supplier approval policy, truck inspection | | No |
| Letters of guarantee from suppliers assuring safe and sanitary packing material | | No |
| Supplier approval policy, truck inspection | | No |
| Supplier approval policy, truck inspection | | No |
| Ingredients that are wet or have been wetted will prevent adequate penetration of cure into animal food | All ingredients must be checked for dryness prior to removal from storage; all ingredients must be free of evidence of having been wet or being wet | Yes |
| SSOP & sanitation policy, allergen control policy, segregated storage / dry storage | | No |
| | | |
| Cooler and freezer temperatures are monitored and controlled to ≤41°F for coolers and ≤0°F (−18°C) for freezers; data recorders or alarm system tie-in are utilized during weekends, holidays, and off-hours for all coolers and freezers, as per written prerequisite program | | No |
| | | |
| | | |
| | | |
| | | |
| | | |

**TABLE E.2** *(continued)*

| Ingredient/Processing Step | Food Hazard | Reasonably Likely to Occur? |
| --- | --- | --- |
| **7. PREPARATION & MEASURING OF NON-MEAT INGREDIENTS** | BIOLOGICAL—Vegetative pathogens such as *E. coli* O157:H7, *Salmonella* ssp., *L. monocytogenes*, *S. aureas*, *T. spiralis* | No |
| | CHEMICAL—Allergens present in facility, sodium nitrite/nitrate | No |
| | PHYSICAL—None | |
| **8. CUTTING, TRIMMING, DE-BONING** | BIOLOGICAL—Vegetative pathogens such as *E. coli* O157:H7, *Salmonella* ssp., *L. monocytogenes*, *S. aureas*, *T. spiralis* | No |
| | CHEMICAL—Allergens present in facility | No |
| | PHYSICAL—Bone fragments and foreign material contamination | No |
| **9. WEIGHING/METERING** | BIOLOGICAL—Vegetative pathogens such as *E. coli* O157:H7, *Salmonella* ssp., *L. monocytogenes*, *S. aureas*, *T. spiralis* | No |
| | CHEMICAL—Allergens present in facility | No |
| | PHYSICAL—None | |
| **10. APPLICATION OF CURE ACCORDING TO RECIPE (CCP 2)** | BIOLOGICAL—Vegetative pathogens such as *E. coli* O157:H7, *Salmonella* ssp., *L. monocytogenes*, *S. aureas*, *T. spiralis* | Yes |
| | CHEMICAL—Allergens present in facility, sodium nitrite/nitrate | No |
| | PHYSICAL—None | |
| **11. VACUUM PACKAGING** | BIOLOGICAL—Vegetative pathogens such as *E. coli* O157:H7, *Salmonella* ssp., *L. monocytogenes*, *S. aureas*, *T. spiralis* | No |

## DRY-CURED ANIMAL FOODS HACCP PLAN

| Basis | If yes to column 3, what measures could be applied to prevent, eliminate, or reduce the hazard to an acceptable level? | Critical Control Point |
|---|---|---|
| Ingredients are prepared and measured per recipe; SSOP & sanitation policy | | No |
| Ingredients are prepared and measured per recipe; SSOP & sanitation policy, allergen control policy, restricted ingredient log | | No |
| | | |
| Animal foods are stored in a temperature-controlled cooler prior to this step, and processed in a temperature-controlled area at ≤ 44°F (7°C); prerequisite programs such as SSOP & sanitation policy in addition to curing and drying step also control hazard | | No |
| SSOP & sanitation program, allergen control policy | | No |
| Animal food is carefully examined through visual inspection to make sure no bone fragments remain and no tools or foreign objects are left behind | | No |
| Animal foods are processed in a temperature-controlled area at ≤ 44°F (7°C); prerequisite programs such as SSOP & sanitation policy in addition to curing and drying step also control hazard | | No |
| SSOP & sanitation policy, allergen control policy | | No |
| | | |
| Improperly measured and prepared ingredients may lead to ineffective or incomplete curing of raw pork | Ingredients must be weighed on a calibrated scale; dry-curing mixture must be prepared according to recipe and applied according to correct ratio of cure to animal food | Yes |
| SSOP & sanitation policy, allergen control policy, restricted ingredient log | | No |
| | | |
| Animal foods are processed in a temperature-controlled area at ≤ 44°F (7°C); pouches are labeled with proper instructions; prerequisite programs such as SSOP & sanitation policy in addition to curing and drying step also control hazard | | No |

**TABLE E.2** *(continued)*

| Ingredient/Processing Step | Food Hazard | Reasonably Likely to Occur? |
|---|---|---|
| | CHEMICAL—Allergens present in facility | No |
| | PHYSICAL—None | |
| **12. COLD HOLDING (≥ 41°F/5°C) (CCP 3)** | BIOLOGICAL—Vegetative pathogens such as *E. coli* O157:H7, *Salmonella* ssp., *L. monocytogenes*, *S. aureas*, *T. spiralis* | Yes |
| | CHEMICAL—Allergens present in facility | No |
| | PHYSICAL—None | |
| **13. PACKAGE REMOVAL & TRANSFER TO OPEN CONTAINER** | BIOLOGICAL—Vegetative pathogens such as *E. coli* O157:H7, *Salmonella* ssp., *L. monocytogenes*, *S. aureas*, *T. spiralis* | No |
| | CHEMICAL—Allergens present in facility | No |
| | PHYSICAL—None | |
| **14. COLD STORAGE FOR RESTING (≤ 41°F/5°C)** | BIOLOGICAL—Vegetative pathogens such as *E. coli* O157:H7, *Salmonella* ssp., *L. monocytogenes*, *S. aureas*, *T. spiralis* | No |
| | CHEMICAL—Allergens present in facility | No |
| | PHYSICAL—None | |
| **15. MOLD INOCULATION** | BIOLOGICAL—Vegetative pathogens such as *E. coli* O157:H7, *Salmonella* ssp., *L. monocytogenes*, *S. aureas*, *T. spiralis* | No |
| | CHEMICAL—Allergens present in facility | No |
| | PHYSICAL—None | |
| **16. INCUBATION (CCP 4)** | BIOLOGICAL—Vegetative pathogens such as *E. coli* O157:H7, *Salmonella* ssp., *L. monocytogenes*, *S. aureas*, *T. spiralis* | Yes |

## DRY-CURED ANIMAL FOODS HACCP PLAN

| Basis | If yes to column 3, what measures could be applied to prevent, eliminate, or reduce the hazard to an acceptable level? | Critical Control Point |
|---|---|---|
| SSOP & sanitation policy, allergen control policy | | No |
| | | |
| Improper time and temperature of holding period may lead to an incomplete curing process | Holding occurs under refrigeration (≤ 41°F/5°C); length of holding time is calculated by 1 day per 0.5 inch (13 mm) of muscle (center/thickest part) in addition to 24 hours | Yes |
| SSOP & sanitation policy, allergen control policy, segregated storage | | No |
| | | |
| Animal foods are processed in a temperature-controlled area at ≤ 44°F (7°C); prerequisite programs such as SSOP & sanitation policy in addition to curing and drying step also control hazard | | No |
| SSOP & sanitation policy, allergen control policy | | No |
| | | |
| Cooler and freezer temperatures are monitored and controlled to ≤ 41°F (5°C) for coolers and ≤0°F (−18°C) for freezers; data recorders or alarm system tie-in are utilized during weekends, holidays, and off-hours for all coolers and freezers, as per written prerequisite program; SSOP & sanitation policy; curing and drying step | | No |
| SSOP & sanitation policy, allergen control policy | | No |
| | | |
| Ingredient mixture for mold inoculation is scaled, prepared, and applied according to recipe | | No |
| SSOP & sanitation policy, allergen control policy | | No |
| | | |
| Cured animal foods incubated in an uncontrolled environment might be at risk for biological hazard | Incubation is monitored and occurs at 85°F (29°C) and 90% RH for 36 to 48 hours | Yes |

**TABLE E.2** *(continued)*

| Ingredient/Processing Step | Food Hazard | Reasonably Likely to Occur? |
|---|---|---|
| | CHEMICAL—Allergens present in facility | No |
| | PHYSICAL—None | |
| **17. WEIGHING** | BIOLOGICAL—Vegetative pathogens such as *E. coli* O157:H7, *Salmonella* ssp., *L. monocytogenes*, *S. aureas*, *T. spiralis* | No |
| | CHEMICAL—Allergens present in facility | No |
| | PHYSICAL—None | |
| **18. DRYING (CCP 5)** | BIOLOGICAL—Vegetative pathogens such as *E. coli* O157:H7, *Salmonella* ssp., *L. monocytogenes*, *S. aureas*, *T. spiralis* | Yes |
| | CHEMICAL—Allergens present in facility | No |
| | PHYSICAL—None | |
| **19. WEIGHING** | BIOLOGICAL—Vegetative pathogens such as *E. coli* O157:H7, *Salmonella* ssp., *L. monocytogenes*, *S. aureas*, *T. spiralis* | No |
| | CHEMICAL—Allergens present in facility | No |
| | PHYSICAL—None | |
| **20. PACKAGING & LABELING** | BIOLOGICAL—Vegetative pathogens such as *E. coli* O157:H7, *Salmonella* ssp., *L. monocytogenes*, *S. aureas*, *T. spiralis* | No |
| | CHEMICAL—Allergens present in facility | No |
| | PHYSICAL—None | |
| **21. COLD STORAGE (≤ 41°F/5°C)** | BIOLOGICAL—Vegetative pathogens such as *E. coli* O157:H7, *Salmonella* ssp., *L. monocytogenes*, *S. aureas*, *T. spiralis* | No |
| | CHEMICAL—Allergens present in facility | No |
| | PHYSICAL—None | |

## DRY-CURED ANIMAL FOODS HACCP PLAN

| Basis | If yes to column 3, what measures could be applied to prevent, eliminate, or reduce the hazard to an acceptable level? | Critical Control Point |
|---|---|---|
| SSOP & sanitation policy, allergen control policy | | No |
| | | |
| SSOP & sanitation policy in addition to curing and drying step also control hazard | | No |
| SSOP & sanitation policy, allergen control policy | | No |
| | | |
| Improper time and temperature of drying period may lead to an incomplete drying process | Drying process occurs at 55°F (13°C) and 85 to 95% RH until weight loss of 35% has been reached; if Cure #1 is used, drying occurs for ≥ 14 days; if Cure #2 is used, drying occurs for ≥ 45 days or longer | Yes |
| SSOP & sanitation policy, allergen control policy | | No |
| | | |
| SSOP & sanitation policy in addition to curing and drying step also control hazard | | No |
| SSOP & sanitation policy, allergen control policy | | No |
| | | |
| SSOP & sanitation policy in addition to curing and drying step also control hazard; product is shelf-stable | | No |
| Allergen control policy, label check | | No |
| | | |
| Cooler and freezer temperatures are monitored and controlled to ≤ 41°F (5°C) for coolers and ≤0°F (−18°C) for freezers; data recorders or alarm system tie-in are utilized during weekends, holidays, and off-hours for all coolers and freezers, as per written prerequisite program | | No |
| Allergen control policy, label check | | No |
| | | |

**TABLE E.2** (*continued*)

| Ingredient/Processing Step | Food Hazard | Reasonably Likely to Occur? |
|---|---|---|
| **22. IN-STORE SALE TO CUSTOMER** | BIOLOGICAL—Vegetative pathogens such as *E. coli* O157:H7, *Salmonella* ssp., *L. monocytogenes*, *S. aureas*, *T. spiralis* | No |
| | CHEMICAL—None | |
| | PHYSICAL—None | |

**TABLE E.3.** Dry-Cured Animal Foods HACCP Plan

| Control Point | Hazard | Critical Limit | Validation |
|---|---|---|---|
| **INGREDIENT DRY STORAGE (CCP 1)** | BIOLOGICAL—Vegetative pathogens such as *E. coli* O157:H7, *Salmonella* ssp., *L. monocytogenes*, *S. aureas*, *T. spiralis* | All ingredients must be checked for dryness prior to removal from storage; all ingredients must be free of evidence of having been wet or being wet | |
| **APPLICATION OF CURE ACCORDING TO RECIPE (CCP 2)** | BIOLOGICAL—Vegetative pathogens such as *E. coli* O157:H7, *Salmonella* ssp., *L. monocytogenes*, *S. aureas*, *T. spiralis* | Ingredients must be weighed on a calibrated scale; standard percentage of cure to meat (100%) is 2.5% salt and 0.25% Cure #1 or Cure #2; sodium nitrite must not exceed 200 ppm and sodium nitrate must not exceed 500 ppm in finished meat product | 21 CFR 172.175 |
| **VACUUM PACKAGING** | BIOLOGICAL—Vegetative pathogens such as *E. coli* O157:H7, *Salmonella* ssp., *L. monocytogenes*, *S. aureas*, *T. spiralis* | Each vacuum-sealed pouch must be labeled with packaging date and instructions to maintain at ≤ 41°F (5°C), and discard within 30 days if not used or consumed | 2016 Ohio Uniform Code §3717-1-03.4 (K)(2)(c) |

| Basis | If yes to column 3, what measures could be applied to prevent, eliminate, or reduce the hazard to an acceptable level? | Critical Control Point |
|---|---|---|
| Labeling instructions to keep items refrigerated or frozen prevent risk of pathogen growth | | No |
| | | |
| | | |

| Monitoring Procedure | Verification | Equipment Used | Corrective Action |
|---|---|---|---|
| Trained staff must visually check curing ingredients for dryness prior to use, monitored by supervisor or food safety lead once per month | Visual inspection | N/A | If deviation is discovered (e.g., ingredient wetness is observed), ingredients may not be used and must be discarded |
| Only trained staff are allowed to prepare and apply cure, monitored once per month by supervisor or food safety lead | Visual inspection; calibration of scale | Calibrated scale, measuring devices | If deviation is discovered during preparation (e.g., incorrect measurement), corrective action may be performed; if deviation is observed after processing step (e.g., after cure has been applied and transferred to cold hold), food must be discarded |
| Only trained staff are allowed to vacuum-pack and label items, monitored once per month by supervisor or food safety lead | Visual inspection; label check | Tape & marker for labeling | If deviation is discovered during process step (e.g., incorrect label), corrective action may be performed; if deviation is observed after processing step (e.g., label not present after storage), food must be discarded |

**TABLE E.3** *(continued)*

| Control Point | Hazard | Critical Limit | Validation |
|---|---|---|---|
| **COLD HOLDING (≤41°F/5°C) (CCP 3)** | BIOLOGICAL— Vegetative pathogens such as *E. coli* O157:H7, *Salmonella* ssp., *L. monocytogenes*, *S. aureas*, *T. spiralis* | Holding occurs under refrigeration (≤ 41°F/5°C); length of holding time is calculated by 1 day per 0.5 inch (13 mm) of muscle (center / thickest part) in addition to 24 hours followed by an uncovered 36-hour rest period | Per Larder Dry-Cure Recipe |
| **MOLD INOCULATION** | BIOLOGICAL— Vegetative pathogens such as *E. coli* O157:H7, *Salmonella* ssp., *L. monocytogenes*, *S. aureas*, *T. spiralis* | *A. oryzae* (5 ml per kg of meat / 0.08 ounce per pound) is mixed with rice flour (120 g per kg of meat / 1.92 ounces per pound) and coated onto animal food | Attachment 1— *Final Risk Assessment of* Aspergillus oryzae (February 1997) |
| **INCUBATION (CCP 4)** | BIOLOGICAL— Vegetative pathogens such as *E. coli* O157:H7, *Salmonella* ssp., *L. monocytogenes*, *S. aureas*, *T. spiralis* | Incubation occurs at 85°F (29°C) and 90% RH for 36 to 48 hours | Attachment 1— *Final Risk Assessment of* Aspergillus oryzae (February 1997); per article, *A. oryzae* is at risk of producing toxin only when fungal incubation exceeds 3 days |
| **DRYING (CCP 5)** | BIOLOGICAL— Vegetative pathogens such as *E. coli* O157:H7, *Salmonella* ssp., *L. monocytogenes*, *S. aureas*, *T. spiralis* | Drying process occurs at 55°F (13°C) and 85 to 95% RH until weight loss of 35% has been reached; if Cure #1 is used, drying occurs for ≥ 14 days; if Cure #2 is used, drying occurs for ≥ 45 days or longer | Per Larder *Dry-Cure Recipe* |

| Monitoring Procedure | Verification | Equipment Used | Corrective Action |
|---|---|---|---|
| Trained staff must visually check temperature of coolers and freezers twice per day | Visual inspection | Digital temperature display; coolers and freezer must also be equipped with an electronic temperature monitoring device | If deviation is discovered (e.g., cooler temperature rises above 41°F/5°C), corrective action may be performed to adjust; temperature of reduced oxygen packaging animal foods must be measured and if below 41°F may remain (if temperature of cooler/freezer has been correctly adjusted) or moved to a properly working cooler/freezer; if temperature increased above 41°F, food must be discarded |
| Only trained staff are allowed to prepare and apply mold, monitored once per month by supervisor or food safety lead | Visual inspection; calibration of scale / use of measuring equipment | Calibrated scale, measuring devices | If deviation is discovered during preparation (e.g., incorrect measurement), corrective action may be performed; if deviation is observed after processing step (e.g., after mold has been applied and transferred to incubation), food must be discarded |
| Trained staff must visually monitor incubation environment | *Dry-Cure Animal Foods Monitoring Log* | Digital temperature display, calibrated RH gauge, time monitoring device | If deviation is discovered during process (e.g., improper temperature), corrective action may be performed (e.g., temperature adjustment); if deviation is observed after processing step (e.g., log was not filled out), food must be discarded |
| Trained staff must visually monitor drying environment | *Dry-Cure Animal Foods Monitoring Log* | Digital temperature display, calibrated RH gauge, time monitoring device, calibrated scale | If deviation is discovered during process (e.g., improper temperature), corrective action may be performed (e.g., temperature adjustment); if deviation is observed after processing step (e.g., log was not filled out), food must be discarded |

# RESOURCES

## CONTRIBUTORS, SUPPORTERS, AND FOLKS WHO KNOW KOJI

This is the listing of all the amazing people who helped to make this book a reality. Without them and their expertise this book would not be what it is. They are the makers, scientists, chefs, brewers, and writers whose passion for koji knows no bounds. They are all responsible for creating fabulous communities around koji across the globe. Please follow along with them on social media, buy their books, visit their shops, and eat at their restaurants.

**Akiko Katayama**
Director, New York Japanese Culinary Academy
@akikokatayamanyc
https://www.akikokatayama.com

**Alfred Francese**
Larder master, Emmer & Rye, Austin, TX
@alfred.fr

**Alex Talbot & Aki Kamozawa**
Authors of Ideas in Food and R&D chef owners, Curiosity Doughnuts, Spring House, PA
@ideasinfood
@curiositydoughnuts
https://blog.ideasinfood.com
http://www.curiositydoughnuts.com

**Anna Markow**
Pastry chef in New York City
@verysmallanna

**Arielle Johnson, PhD**
Science officer on Alton Brown's show *Good Eats*
@arielle_johnson

**Bentley Lim, PhD**
Microbiology researcher, Yale University, New Haven, CT
https://www.researchgate.net/publication/316309161_Engineered_Regulatory_Systems_Modulate_Gene_Expression_of_Human_Commensals_in_the_Gut

**Branden Byers**
Author of *The Everyday Fermentation Handbook* and host of *FermUp*
@fermup
http://fermup.com

**Brian Benchek**
Brewmaster-owner, Bottlehouse Brewery & Mead Hall, Cleveland, OH
@bottlehousebrew
@thebottlehousebrewery
https://www.bottlehouse.co

**Brian & Mickey Kellerman**
Food safety consultants and HACCP specialists, Kellerman Consulting, Columbus, OH
@kellermanohio
@kellermanconsulting
https://www.kellermanconsulting.com

**Coral Lee**
Gastronome, journalist, and host of *Meant to Be Eaten*
@meanttobeeaten
https://heritageradionetwork.org/series/meant-to-be-eaten

**Kevin Farley & Alex Hozven**
Cultured Pickle Shop, Berkeley, CA
@culturedpickleshop
https://www.culturedpickleshop.com

**Cynthia Graber**
Journalist, cohost of *Gastropod*
@cagraber
@gastropodcast
https://gastropod.com

**David Uygur**
Chef-owner and Prince of Pork, Lucia, Dallas, TX
@lucia_dallas
https://www.luciadallas.com/#

**Diana Clark**
Bovine anatomist, Certified Angus Beef, Wooster, OH
@certifiedangusbeef
https://www.certifiedangusbeef.com

**Eric Edgin**
Woodworker
@ericedgin

**Eugene Zeleny**
Engineer and woodworker, MIT Media Lab, Cambridge, MA, and WouldWork Shop, Brighton, MA
@wouldwork_shop
https://www.etsy.com/shop/WouldWorkShop

**Irene Yoo**
Zymologist and chef, Yoo Eating, New York, NY
@yooeating
@heyireneyoo
http://www.yooeating.com

**James Wayman**
Chef, Oyster Club, Engine Room, and Grass & Bone, Mystic, CT
@jameswayman
http://www.oysterclubct.com

**Jeremy Kean**
Chef-owner, Brassica Kitchen + Café, Jamaica Plain, MA
@brassicakitchen
@keanjeremy
https://www.brassicakitchen.com

**Jon Adler**
Sommelier, SingleThread Farm Restaurant Inn, Healdsburg, CA
@jonadzer

**John Gibbons, PhD**
Evolutionary geonomics, Clark University, Worcester, MA

**Johnny Drain**
Head of Fermentation R&D, Cub in London and coeditor of MOLD magazine
@drjohnnydrain

**John Hutt**
Chef and researcher, Museum of Food and Drink, Brooklyn, NY
@crasstafarian
@mofad
@MOFADInfo
https://www.mofad.org

## RESOURCES

**Josh Evans**
Food researcher
www.joshuadavidevans.com

**Harry Rosenblum**
Founder, The Brooklyn Kitchen, Brooklyn, NY, and author of *Vinegar Revival*
@thebklynkitchen
https://www.thebrooklynkitchen.com

**Ken Fornataro**
Chef/Zymologist, Cultures Group, New York, NY
@culturesgroup
https://cultures.group

**Kevin Fink**
Chef-owner, Emmer & Rye, Austin, TX
@emmerandrye
https://emmerandrye.com

**Kim Wejendorp**
Head of R&D at Amass Restaurant, Copenhagen, Denmark
@kimwejendorp
https://amassrestaurant.com/

**Kirsten Shockey**
Author of *Fermented Vegetables*, *Fiery Ferments*, and *Miso, Tempeh, Natto and Other Tasty Ferments*
@kirstenstockey
http://ferment.works

**Koichi Higuchi**
Seventh-generation koji spore producer, Higuchi Matsunosuke Shoten Co. Ltd., Osaka, Japan
http://www.higuchi-m.co.jp/english/index.html

**Kristyn & Kevin Henslee**
Farmers, cheesemakers, and owners, Yellow House Cheese, Seville, OH
@yellowhousecheese
@yhc_yellowhouse
http://www.yellowhousecheese.com

**Mara Jane King**
Zymologist and owner, Ozuke, Boulder, CO
@zukemono
@we.are.ozuke
@weareozuke
https://ozuke.com

**Meredith Leigh**
Author of *The Ethical Meat Handbook* and *Pure Charcuterie*
@mereleighfood
http://www.mereleighfood.com

**Misti Norris**
Chef-owner, Petra and the Beast, TX
@misti.j.norris
@petraandthebeast

**Nicco Muratore**
Chef de cuisine, Commonwealth Cambridge, Cambridge, MA
@niccomuratore

**Sam Jett**
Head of R&D and partner, Patchwork Productions, Nashville, TN
@samuel.jett

**Sean Doherty**
Chef and baker in Portland, ME
@naes2020

**Stephen Lyman**
Shochu ambassador and author of *The Complete Guide to Japanese Drinks*; Shochu educator, Sake School of America
@shochu_danji
http://kampai.us

**Sarah Conezio & Isaiah Billington**
Owners, White Rose Miso and Keepwell Vinegar, Maryland
@whiterosemiso
@keepwellvinegar
https://www.keepwellvinegar.com

## BOOKS

*The Art of Fermentation* by Sandor Ellix Katz (Chelsea Green Publishing, 2012)
*The Book of Miso* by William Shurtleff and Akiko Aoyagi (CreateSpace, 2018)
*Miso, Tempeh, Natto and Other Tasty Ferments* by Kirsten and Christopher Shockey (Storey, 2019)
*The Noma Guide to Fermentation* by René Redzepi and David Zilber (Artisan, 2018)
*Preserving the Japanese Way* by Nancy Singleton Hachisu (Andrews McMeel Publishing, 2015)
*On Food and Cooking* by Harold McGee (Scribner, revised and updated 2004)
*The Oxford Companion to Food* by Alan Davidson (Oxford University Press, third edition 2014)

# NOTES

## CHAPTER 1: WHAT IS KOJI?


1. Patrick E. McGovern et al., "Fermented Beverages of Pre- and Proto-Historic China," *Proceedings of the National Academy of Sciences of the United States of America* 101, no. 51 (December 2004): 17593–98, doi:10.1073/pnas.0407921102.
2. Masayuki Machida, Osamu Yamada, and Katsuya Gomi, "Genomics of *Aspergillus oryzae*: Learning from the History of Koji Mold and Exploration of Its Future," *DNA Research* 15, no. 4 (August 2008): 173–83, doi:10.1093/dnares/dsn020.
3. Kathleen Tuthill, "John Snow and the Broad Street Pump on the Trail of an Epidemic," *Cricket Magazine* 31, no. 3 (November 2003), https://www.ph.ucla.edu/epi/snow/snowcricketarticle.html.
4. McGovern et al., "Fermented Beverages."
5. William Shurtleff and Akiko Aoyagi, *History of Miso, Soybean Jiang (China), Jang (Korea), and Tauco / Taotjo (Indonesia) (200 BC–2009)* (Lafayette, CA: Soyinfo Center, 2009), 7.
6. Tove Christensen et al., "High Level Expression of Recombinant Genes in *Aspergillus oryzae*," *Nature Biotechnology* 6 (December 1988): 1419–22.
7. Justin P. Jahnke et al., "*Aspergillus oryzae–Saccharomyces cerevisiae* Consortium Allows Bio-Hybrid Fuel Cell to Run on Complex Carbohydrates," *Microorganisms* 4, no. 1 (February 2016): 10, doi: 10.3390/microorganisms4010010; Hiroshi Maeda et al., "Purification and Characterization of a Biodegradable Plastic-Degrading Enzyme from *Aspergillus oryzae*," *Applied Microbiology Biotechnology* 67, no. 6 (June 2005): 778–88, doi: 10.1007/s00253-004-1853-6.
8. Hiroshi Hamajima et al., "Japanese Traditional Dietary Fungus Koji *Aspergillus oryzae* Functions as a Prebiotic for *Blautia coccoides* Through Glycosylceramide: Japanese Dietary Fungus Koji Is a New Prebiotic," *Springerplus* 5, no. 1 (August 2016): 1321, doi:10.1186/s40064-016-2950-6.


## CHAPTER 3: THE FLAVOR-MAKING ROAD MAP


1. Harold McGee, *On Food and Cooking: The Science and Lore of the Kitchen* (New York: Scribner, 2004), 155–56.


## CHAPTER 5: EXPANDING YOUR KOJI MAKING


1. D.K. O'Toole, "Soybean: Soy-Based Fermented Foods," in *Encyclopedia of Grain Science* ed. Colin W. Wrigley, Harold Corke, and Charles Walker (Amsterdam: Elsevier, 2004), 180, doi:10.1016/b978-0-12-394437-5.00129-7.
2. Y. H. Yui et al., *Handbook of Food and Beverage Fermentation Technology* (New York: Marcel Dekker, 2004), 503.
3. Yui et al., 545.

### CHAPTER 6: SHORT-TERM ENHANCEMENT

1. Yoshifumi Oguro, Ayana Nakamura, and Atsushi Kurahashi, "Effect of temperature on saccharification and oligosaccharide production efficiency in koji amazake," *Journal of Bioscience and Bioengineering* 127, no. 5 (2019): 570-574, doi:10.1016/j.jbiosc.2018.10.007.
2. Ruann Janser Soares de Castro and Helia Harumi Sato, "Protease from *Aspergillus oryzae*: Biochemical Characterization and Application as a Potential Biocatalyst for Production of Protein Hydrolysates with Antioxidant Activities," Journal of Food Processing (2014): Article ID 372352, doi:10.1155/2014/372352.
3. William Shurtleff and Akiko Aoyagi, *The Book of Miso* (CreateSpace, 2018), 162.
4. Shurtleff and Aoyagi, *The Book of Miso*

### CHAPTER 7: AMINO PASTES

1. Nancy Singleton Hachisu, *Preserving the Japanese Way: Traditions of Salting, Fermenting, and Pickling for the Modern Kitchen* (Kansas City, MI: Andrews McMeel Publishing, 2015), 52.
2. Stan Kubow, "Routes of Formation and Toxic Consequences of Lipid Oxidation Products in Foods," *Free Radical Biology and Medicine* 12, no. 1 (1992): 63–81, doi:10.1016/0891-5849(92)90059-P; Samantha A. Vieira, David Julian McClements, and Eric A. Decker, "Challenges of Utilizing Healthy Fats in Foods," *Advances in Nutrition* 6, no. 3 (May 7, 2015): 309S–17S, doi:10.3945/an.114.006965.
3. P. L. Pavcek and G. M. Shull, "Inactivation of Biotin by Rancid Fats," *Journal of Biological Chemistry* 146, no. 2 (December 1942): 351–55.
4. Bing-Jian Feng et al., "Dietary Risk Factors for Nasopharyngeal Carcinoma in Maghrebian Countries," *International Journal of Cancer* 121, no. 7 (June 2007): 1550–55, doi:10.1002/ijc.22813; Ying-Chin Ko et al., "Chinese Food Cooking and Lung Cancer in Women Nonsmokers," *American Journal of Epidemiology* 151, no. 2 (January 2000): 140–47, doi:10.1093/oxfordjournals.aje.a010181.
5. Tom P. Coulate, *Food: The Chemistry of Its Components* (Cambridge, U.K.: Royal Society of Chemistry, 2009), 122–23.
6. McGee, *On Food and Cooking*, 145.
7. KeShun Liu, "Food Use of Whole Soybeans," in *Soybeans: Chemistry, Production, Processing, and Utilization*, ed. Lawrence A. Johnson, Pamela J. White, and Richard Galloway (Urbana, IL: AOCS Press, 2008): 441–481, doi:10.1016/B978-1-893997-64-6.50017-2.
8. William Shurtleff and Akiko Aoyagi, "A Comprehensive History of Soy" (Lafayette, CA: Soyinfo Center, 2019), http://www.soyinfocenter.com/HSS.
9. Seung-Beom Hong, Dae-Ho Kim, and Robert A. Samson, "*Aspergillus* Associated with Meju, a Fermented Soybean Starting Material for Traditional Soy Sauce and Soybean Paste in Korea," *Mycobiology* 43, no. 3 (2015): 218–24, doi:10.5941/myco.2015.43.3.218; Dae-Ho Kim et al. "Mycoflora of Soybeans Used for Meju Fermentation," *Mycobiology* 41, no. 2 (2013): 100–07, doi:10.5941/myco.2013.41.2.100.
10. Hyeong-Eun Kim, Song-Yi Han, and Yong-Suk Kim, "Quality Characteristics of Gochujang Meju Prepared with Different

Fermentation Tools and Inoculation Time of *Aspergillus oryzae*," *Food Science and Biotechnology* 19, no. 6 (December 2010): 1579–85, doi:10.1007/s10068-010-0224-6.
11. D. Y. Kwon et al., "Isoflavonoids and Peptides from Meju, Long-Term Fermented Soybeans, Increase Insulin Sensitivity and Exert Insulinotropic Effects in Vitro," *Nutrition* 27, no. 2 (February 2011): 244–52, doi:10.1016/j.nut.2010.02.004.
12. Su Yun Lee et al., "Mass Spectrometry-Based Metabolite Profiling and Bacterial Diversity Characterization of Korean Traditional Meju During Fermentation," *Journal of Microbiology and Biotechnology* 22, no. 11 (November 2012): 1523–31, doi:10.4014/jmb.1207.07003.
13. Dae-Ho Kim et al., "Fungal Diversity of Rice Straw for Meju Fermentation," *Journal of Microbiology and Biotechnology* 23, no. 12 (November 2013): 1654–63, doi:10.4014/jmb.1307.07071.
14. Kim et al., "Mycoflora of Soybeans Used for Meju Fermentation."
15. Woo Yong Jung et al., "Functional Characterization of Bacterial Communities Responsible for Fermentation of Doenjang: A Traditional Korean Fermented Soybean Paste," *Frontiers in Microbiology* 7 (May 2016): 827, doi:10.3389/fmicb.2016.00827.

## CHAPTER 8: AMINO SAUCES

1. McGee, *On Food and Cooking.*
2. Jennifer LeMesurier, "Uptaking Race: Genre, MSG, and Chinese Dinner," *Poroi: An Interdisciplinary Journal of Rhetorical Analysis and Invention* 12, no. 2 (2017): Article 7, doi:10.13008/2151-2957.1253.
3. William Shurtleff and Akiko Aoyagi, "History of Soy Sauce, Shoyu, and Tamari" (Lafayette, CA: Soyinfo Center, 2004), 4, http://www.soyinfocenter.com/HSS/soy_sauce4.php.
4. Shurtleff and Aoyagi, "History of Soy Sauce, Shoyu, and Tamari," 4.
5. Yui et al., *Handbook of Food and Beverage Fermentation Technology*, 507.

## CHAPTER 9: ALCOHOL AND VINEGAR

1. Lisa Solieri et al., *Vinegars of the World* (Milan: Springer, 2009), 22–23.

## CHAPTER 10: AGING MEAT AND CHARCUTERIE

1. 5m Editor, "Ageing and the Impact on Meat Quality," The Pig Site, last modified June 1, 2012, https://thepigsite.com/articles/ageing-and-the-impact-on-meat-quality.
2. Sue Shephard, *Pickled, Potted, and Canned: How the Art and Science of Food Preserving Changed the World* (New York: Simon & Schuster, 2000), 70.
3. Marta Laranjo, Miguel Elias, and Maria João Fraqueza, "The Use of Starter Cultures in Traditional Meat Products," *Journal of Food Quality* 3 (2017): 1–18, doi:10.1155/2017/9546026.

## CHAPTER 12: VEGETABLES

1. William Shurtleff and Akiko Aoyagi, *The Book of Tofu and Miso* (Berkeley: Ten Speed Press, 2001), 41.
2. Yunping Zhu, "Characterization of a Naringinase from *Aspergillus oryzae* 11250 and Its Application in the Debitterization of Orange Juice," *Process Biochemistry* 62 (November 2017): 114–21.

## APPENDIX B: A DEEPER DIVE INTO MAKING AMINOS

1. Toshihide Nishimura and Hiromichi Kato, "Taste of Free Amino Acids and Peptides," *Food Reviews International* 4, no. 2 (January 1988): 175–94, doi:10.1080/87559128809540828.
2. Ninomiya Yuzo et al., "Taste Synergism Between Monosodium Glutamate and 5'-Ribonucleotide in Mice," *Comparative Biochemistry and Physiology Part A: Physiology* 101, no. 1 (January 1992): 97–102, doi:10.1016/0300-9629(92)90634-3.
3. Juerg Solms, "Taste of Amino Acids, Peptides, and Proteins," *Journal of Agricultural and Food Chemistry* 17, no. 4 (July 1969): 686–88, doi:10.1021/jf60164a016.
4. Cindy J. Zhao, Andreas Schieber, and Michael G. Gänzle, "Formation of Taste-Active Amino Acids, Amino Acid Derivatives and Peptides in Food Fermentations—A Review," *Food Research International* 89, no. 1 (August 2016): 39–47, doi:10.1016/j.foodres.2016.08.042.
5. Yang Yang et al., "Dynamics of Microbial Community During the Extremely Long-Term Fermentation Process of a Traditional Soy Sauce," *Journal of the Science of Food and Agriculture* 97, no. 10 (February 13, 2017): 3220–27, doi:10.1002/jsfa.8169.
6. Soichi Furukawa et al., "Significance of Microbial Symbiotic Coexistence in Traditional Fermentation," *Journal of Bioscience and Bioengineering* 116, no. 5 (November 2013): 533–39, doi:10.1016/j.jbiosc.2013.05.017.
7. Xiaohong Cao et al., "Genome Shuffling of *Zygosaccharomyces rouxii* to Accelerate and Enhance the Flavour Formation of Soy Sauce," *Journal of the Science of Food and Agriculture* 90, no. 2 (November 2009): 281–85, doi:10.1002/jsfa.3810; Xiaohong Cao et al., "Improvement of Soy-Sauce Flavour by Genome Shuffling in *Candida versatilis* to Improve Salt Stress Resistance," *International Journal of Food Science & Technology* 45, no. 1 (December 11, 2009): 17–22, doi:10.1111/j.1365-2621.2009.02085.x.
8. Yutaka Maruyama, "Umami Responses in Mouse Taste Cells Indicate More than One Receptor," *Journal of Neuroscience* 26, no. 8 (February 22, 2006): 2227–34, doi:10.1523/JNEUROSCI.4329-05.2006.
9. Feng Zhang et al., "Molecular Mechanism for the Umami Taste Synergism," *Proceedings of the National Academy of Sciences of the United States of America* 105, no. 52 (December 2008): 20930–34, doi:10.1073/pnas.0810174106.
10. Jaewon Shim et al., "Modulation of Sweet Taste by Umami Compounds via Sweet Taste Receptor Subunit hT1R2," *PloS One* 10, no. 4 (April 8, 2015): e0124030–39, doi:10.1371/journal.pone.0124030; Min Jung Kim et al., "Umami-Bitter Interactions: The Suppression of Bitterness by Umami Peptides via Human Bitter Taste Receptor," *Biochemical and Biophysical Research Communications* 456, no. 2 (January 9, 2015): 586–90, doi:10.1016/j.bbrc.2014.11.114.
11. D. Glenn Black and Jeffrey T. Barach, *Canned Foods: Principles of Thermal Process Control, Acidification and Container Closure Evaluation*, 8th ed. (Arlington, VA: GMA Science and Education Foundation, 2015).

# INDEX

Note: Page numbers in **bold** refer to recipes. Page numbers in *italics* refer to figures and photographs. Page numbers followed by *t* refer to tables. Page numbers preceded by "ci" refer to the color insert section.

# INDEX

# INDEX

# INDEX

# INDEX

# ABOUT THE AUTHORS

*Catherine Dzilenski, Idlewild Photo*

**Rich Shih** is one of the leading culinary explorers of koji and miso in the United States and an in-demand food preservation consultant, helping chefs to build their larders and leverage fermentation to decrease waste, and offering ideas with which to experiment. Shih offers both public and private workshops across the United States to share koji knowledge. In addition to working with koji and fermentation, Shih is the exhibit engineer for the Museum of Food and Drink (MOFAD) based in New York City. His blog, *OurCookQuest*, provides a welcome environment for cooks of all experience levels to learn, share knowledge, and exchange ideas.

*Angelo Merendino*

**Jeremy Umansky**, along with Allie La Valle-Umansky and Kenny Scott, is a chef-owner of Larder: A Curated Delicatessen & Bakery in Cleveland, Ohio. A from-scratch Eastern European delicatessen focusing on the use and promotion of local and wild foods, it was nominated by the James Beard Foundation as the Best New Restaurant in America in 2019. Umansky is a foremost expert on koji and fermented, preserved, and foraged foods and works as a consultant on the use and creation of these foods and ingredients.

Articles written by and about Umansky's work have appeared in *Food & Wine*, *Bon Appétit*, and *Saveur*, among other outlets.